HISTORIC RANCHES
of the
Old West

BILL
O'NEAL

EAKIN PRESS ★ Austin, Texas

For my uncle
Ted Standard —
rodeo cowboy and master saddlemaker.

FIRST EDITION

Copyright © 1997
By Bill O'Neal

Published in the United States of America
By Eakin Press
An Imprint of Sunbelt Media, Inc.
P.O. Drawer 90159   ★   Austin, TX 78709-0159

2   3   4   5   6   7   8   9

ISBN 1-57168-167-1

Library of Congress Cataloging-in-Publication Data

O'Neal, Bill, 1942-
    Historic ranches of the Old West / Bill O'Neal.
        p.   cm.
    Includes bibliographical references (p.   ) and index.
    ISBN 1-57168-167-1
    1. Ranches — Southwest, New— History. 2. Ranchers — Southwest, New —
    Biography. 3. Ranch life — Southwest, New — History. 4. Southwest, New —
    History, Local.   I. Title.
    F786.063   1997   97-15250
    979 — dc21   CIP

Contents

Acknowledgments

I first broached the subject of a book about historic western ranches to John Joerschke, the longtime editor of *True West, Old West,* and Barbed Wire Press. John offered immediate encouragement, and throughout this project I have benefited enormously from his innovative suggestions and impressive editorial gifts. I am fortunate to claim John as friend and editor.

In West Texas my wife and I were received with traditional western hospitality: at the Pitchfork by office manager Stella Carter; at the Matador by cowboys Bob Davis and Howard Martin, and retired manager Johnny V. Stevens; at the Slaughter Ranches by manager Jim Bell; at the Spade by manager Donald Vest and his relatives Essie Jourdan and Roger and Jeremy Jeffcoat; at the SMS offices in Stamford by Carl Willis; and at the Running R near Bandera by our informative wrangler, Bob Branham. In Channing we were treated to a personal tour of the XIT general office by W. P. Kirkeminde, a retired veterinarian who has devoted nearly a decade to restoring the historic brick structure. I was invited to the JAs by manager Terry Hawkins and by owner Mary Bivins, who has handsomely maintained John Adair's rambling old house; but when we arrived all adults were away fighting a prairie fire, and we were toured about by Josh Hawkins, twelve-year-old son of the manager and a future top hand.

Near Roswell, New Mexico, we were hosted by Morgan and Joyce Nelson, whose home is on property formerly owned by Pitser Chisum. Morgan took us around John Chisum's famous old ranch headquarters, then shared his extensive files with us. Nora Hand and Bob Hart of Lincoln also offered information about Chisum and his ranch. Through the auspices of New Mexico park rangers Howard Thomas and Orlando Navarrette, of the Oliver Lee State Memorial Park, we were treated to a personal tour of Lee's Dog Canyon Ranch house.

At Arizona's Sierra Bonita Ranch, foreman Les Shannon guided us through the historic headquarters, amiably posing for photos while allowing us to tap his prodigious fund of knowledge about the spread where he has worked most of his life. Max Witkind, Bureau of Land Management archaeologist stationed in Tucson, opened to us the rich materials he has collected on the Empire Ranch. Garnette Franklin of Holbrook, director of the Navaho County Historical Society, shared her files and personal knowledge of the Hash Knife.

Noted Tombstone historian Ben Traywick regaled us with tales of the region and generously loaned me rare books from his personal library. Similar materials also were obtained from another treasured friend, Dr. Phil Earle, director of the Nevada State Historical Society. Still another generous friend, Jim Browning of Charleston, South Carolina, who has traveled to every corner of the West, gave me photos and directions and advice, and even loaned me his camera at the old Two Bar headquarters in Brown's Park, Colorado.

Information about the TA and other Johnson County ranches was gleaned from Gary C. "Andy" Anderson, curator of the Jim Gatchell Memorial Museum in Buffalo, Wyoming. At the Wyoming Hereford Ranch, southeast of Cheyenne, manager Paul Ferguson graciously gave us permission to investigate every structure at the extensive headquarters complex.

Lester Jens, lifelong resident of Terry, Montana, shared his expertise about the northern operations of the XIT. At the Range Rider Museum in Miles City, Bunny Barthelmess Miller provided considerable information about Montana ranches of the area. Again and again we were aided and advised by westerners, several of whom modestly insisted that their names not be mentioned.

My wife, Karon, accompanied me across the West, thousands of miles by plane, car, and even horseback, as we sought out scores of ranches, some abandoned but many still in operation. She shot hundreds of photos, helped me photocopy materials and track down books in numerous libraries, took notes during various interviews, shared the driving, and typed the final manuscript. I could not have written this book without her aid and companionship.

What They Said About Ranching

"Buy land and never sell." — *Richard King*

"These steers are walking fifty-dollar bills." — *Clem Rogers*

"A man who doesn't admire a good beef steer, a good horse, and a pretty woman . . . well, something is wrong with that man's head." — *Tom Waggoner*

"The XIT was noted for good horses and pretty ones, and also for ugly cowpunchers, of which I am a fair sample with plenty of company." — *Alec Sevier, XIT cowpuncher*

"If he stole it to eat, tell him to enjoy it and bring me the hide. If he stole it to sell, bring me his hide." — *Col. E. P. Hardesty of Nevada to his cowboys regarding anyone caught killing his cattle.*

Cowboys around an LIT chuckwagon. The brand was founded by cattleman-banker George W. Littlefield, who eventually sold out, then began again with the LFD brand.
— Courtesy Panhandle-Plains Historical Museum, Canyon

"Sometimes I say that any cowman of open range days who claimed never to have put his brand on somebody else's animal was either a liar or a poor roper." — *Ike Pryor*

"When you can see the cow chips floating, then we've had a rain." — *Sam Ragland, King Ranch livestock manager*

"Not a great deal to say about the trail. In fact, I think it is the easiest part of cow punching . . . Of course we had some storms and some bad nights, but the trail life is easy compared to ranch life." — *Gene Elliston, XIT cowboy*

"Was it civilized in those days? Well, it didn't seem so wild and woolly to us then, but if anybody now had to go back to living like they did then, I reckon about half of us couldn't stand it." — *Weasley Stevens, King Ranch cowboy*

"Loco is a weed that comes early in the spring before the grass comes up. Stock will eat it and the more that eat the more they want. They will hunt it like a dope fiend man will hunt morphine. If they don't get too bad they sometimes get over it and get fat but they never get active any more." — *J.A. Smiley, XIT cowboy*

"I have only one rule in business: When everybody is wanting to sell, I buy; when everybody is wanting to buy, I sell." — *George W. Littlefield*

"I consider twenty-four hours a working day for me. If I get my work done before the twenty-four hours is over, I sleep. But the work must be done, whether I get my rest or not." — *Henry Miller*

"We carry on our business without a ledger of any kind; don't even have a bookkeeper. In fact, I keep my accounts in my head." — *John Sparks, who owned several ranches and 90,000 cattle*

"Save your grass: it is your cheapest and best feed: save it by stocking lightly; use young bulls; subdivide your range if possible; and breed intensively." — *Charles M. O'Donel, Bell Ranch manager*

Introduction

Andy Adams, in his celebrated *The Log of a Cowboy*, related that "as the boys of our family grew old enough the fascination of a horse and saddle was too strong to be resisted." Following his two older brothers into ranch work, "I took to the range as a preacher's son takes to vice." Another teenaged Texan, F. M. Polk, rode on his first trail drive to Kansas at the age of eighteen in 1872: "We were a care-free bunch, had lots of fun and also lots of hard work." Polk's father persuaded him to return to the family farm, "but I did not like farming, and after two years' trial of it, I was more than ready to go back to the wild, care-free life of a cowboy."

This "wild, care-free life" appealed to thousands of other farmers' sons after the Civil War. The western cowboy, riding a swift cow pony and wearing a big hat, spurs, and leather chaps, twirling a lariat while galloping after wild longhorn cattle, swaggering along the boardwalks of Abilene and Dodge City and Tascosa, captivated the public imagination and eventually became the world's premier folk hero.

The central workplace of the cowboy was the ranch, a frontier institution which rapidly acquired its own magnetic appeal. During the formative period of the western cattle industry enormous ranches grew up, encompassing hundreds of thousands — sometimes millions — of acres of open range. The largest of these gigantic spreads employed a hundred or more cowboys tending over 100,000 head of cattle, and the public was fascinated by powerful cattle barons such as Richard King, Charles Goodnight, John Chisum, Henry Hooker, John Sparks, Henry Miller, Burke Burnett, and Shanghai Pierce. Such men controlled vast expanses of wild, magnificent rangelands; they accumulated colossal herds of cattle, maintaining their property through courage, hard work, and sheer force of character. The most successful of the early ranchers proudly ran their cattle kingdoms like feudal nobles, with brands and ear marks their heraldry, hard-riding cowboys their knights, and roundups their tournaments.

Cattle and horses were introduced to the western hemisphere by the Spaniards. *Vaqueros* developed the techniques and specialized equipment for handling range cattle from horseback, and great *haciendas* were established in northern Mexico with thousands of longhorns (a *hacienda* was a large agricultural estate, while a *rancho* was a smaller unit). Spanish missions in Texas and California, as well as to a lesser degree in Arizona and New Mexico, often maintained sizable cattle herds. Californios and, far more profoundly, Texans were influenced by Spanish cattle, horses, and the techniques and equipment used by *vaqueros*.

Immense herds of longhorns and mustangs bred naturally in sparsely settled Texas. By the middle of the nineteenth century Texans began to trail longhorns to various markets, and after the Civil War the famous cattle drives to Kansas railheads captured the nation's attention. Soon Texas herds were placed on the open-range in Kansas, nearer to market railheads, then in Nebraska and Colorado. With the elimination of buffalo herds and the warlike Indian tribes who depended upon them for subsistence, open-range ranchers brought large numbers of cattle into Wyoming, Montana, Dakota, New Mexico, and Arizona. This movement was accelerated by the "Beef Bonanza," as Eastern and British investors poured capital into cattle ranching, hoping to duplicate the success of a few pioneer cattlemen. Another segment of the Great Plains, the reservation lands of Indian Territory, soon were leased by enterprising cattlemen in adjacent states. Like the Great Plains, the desert ranges of the Great Basin in northern Nevada and southern Oregon and Idaho were occupied by cattle ranchers.

Within two decades after the Civil War, the Great Plains and Great Basin had been flooded with cattle herds, cowboys, and sprawling ranches. During the formative period of western cattle ranching, most operations were conducted on open range appropriated by the rancher. For years the grass was free and longhorns were cheap, encouraging the boldest cattlemen to expand their ranches to mythic proportions.

Richard King, whose South Texas ranch would become world-famous, scrupulously acquired title to every parcel of his expanding property. But King was almost unique among early ranchers in his determination to purchase large amounts of land. After the Homestead Act was enacted by Congress in 1862, many ranchers filed a 160-acre claim encompassing a key water source and thereby giving effective control to a large surrounding area of free range. Indeed, many big ranches were built by the legal acquisition of several water sources, which kept anyone else from using the unclaimed but waterless outlying ranges.

If a homestead claim had been obtained, it was here that ranch buildings and corrals were erected. The ranch house might be merely a one-room log cabin or *jacal*, even on a large ranch if the owners were absentee investors. Of course, absentee owners often liked to venture out to their western ranches and frequently built impressive structures for their infrequent visits. There were luxurious, if rustic, hunting lodges, handsome adobes, and towering Victorian structures, many of which still stand.

Near the ranch house was a bunkhouse, cookhouse, cook's cabin, and, often, a cast-iron dinner bell. The headquarters complex also included a horse barn, corrals, sheds, dairy barn, blacksmith shop, harness shop, chicken house, smokehouse, laundry, root cellar, ice house, and outhouses. At some large, isolated ranches, there might even be a post office and a ranch store. Visitors today may tour historic complexes of this size at Montana's Grant-Kohrs Ranch and at John Slaughter's San Bernardino Ranch in southern Arizona.

Life on these ranches was vigorous and exhilarating. Admittedly, existence on small, hardscrabble spreads was brutally monotonous and narrow, but outfits with thousands or tens of thousands of cattle radiated success and excitement. There was a festive, circus-like atmosphere at big roundups, and large crews felt a strong sense of camaraderie and loyalty to "their" ranch. "We thought it was a great honor to work some big cow outfits," reminisced New Mexico cowboy Joe Pankey. "We did what they wanted done, we didn't ask no questions."

Proud of their horsemanship and roping skills, cowboys roamed open country on strong mounts. "Of course, I liked to ride," declared Pankey on behalf of legions of cowboys. "It was about the only thing we could do to enjoy ourselves. That and work."

The work was hard and dirty, but the hard-riding cowboy seized the American consciousness. And he rode the open range of our most colorful frontier, working for lordly men who built epic cattle kingdoms.

Today there are substantial vestiges of a great many of the West's most historic ranches. A number of the old spreads remain in operation, still utilizing the brands and ranch buildings from the nineteenth century, or at least functioning as guest ranches. At other sites the buildings and corrals long have been abandoned, but these venerable log or rock structures exert the same lure for western travelers as a ghost town or decaying cavalry fort. The land, of course, is still there. Much of it is now fenced or permanently changed because of overgrazing, but there are towering mountains and sweeping prairies and open vistas that remain as seen by ranchers and cowboys of a century ago. A surprising number of historic ranch headquarter complexes are maintained by federal, state, and even county or local agencies. The tangible remains of the western ranching industry are substantial, if generally overlooked by travelers, and offer special delights to modern adventurers.

The ranches portrayed in this book represent an arbitrary selection by the author. Many choices were obvious: the King Ranch, perhaps the most famous of all ranches; Charles Goodnight's JAs; Arizona's celebrated Hash Knife; the XIT, more than three million acres of Texas rangeland deeded to a Chicago syndicate willing to erect a massive new capitol building in Austin; and John Chisum's enormous New Mexico empire. In addition to such gargantuan domains, more modest ranches were included because of historical connections: the TA in Wyoming, for example, was a major battle site of the infamous Johnson County War, while Pete Kitchen's Arizona adobe withstood one Apache assault after another. Many celebrities were associated with ranches: Buffalo Bill Cody built his Scout's Rest Ranch into a Nebraska showplace; Theodore Roosevelt and the almost equally dynamic Marquis de Mores were attracted to Dakota Territory during the same period; Will Rogers was born and raised on the historic spread carved by his father from Cherokee lands; and Owen Wister immortalized the Goose Egg in his landmark novel, *The Virginian.*

While various large or important ranches may have been omitted, this book is not intended to be an encyclopedia. An army of researchers would be required to assemble an encyclopedic compilation of sizable and significant western ranches, and readers would find such a multi-volume work imposing and tedious. The purpose of the author has been to concentrate on the most notable ranches of this singularly romantic era — the range cattle industry of the last West.

RANCHING IN TEXAS

Other states were carved or born,
Texas grew from hide and horn.

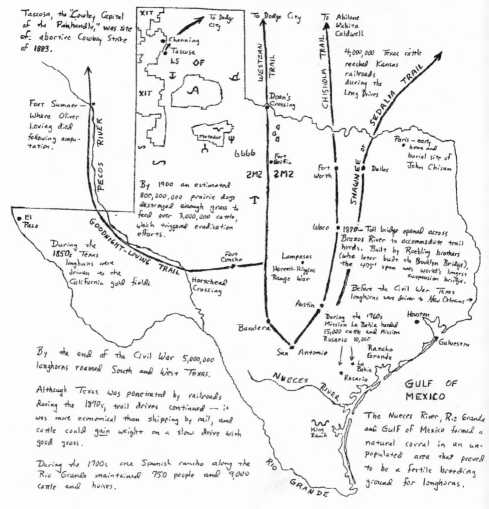

Tascosa, the "Cowboy Capital of the Panhandle," was site of abortive Cowboy Strike of 1883.

Fort Sumner — Where Oliver Loving died following amputation.

XIT

To Dodge City

To Dodge City

To Abilene
Wichita
Caldwell

Channing
Tascosa
LS OF
J
XIT
A

WESTERN TRAIL

CHISHOLM TRAIL

SEDALIA TRAIL

4,000,000 Texas cattle reached Kansas railroads during the Long Drives

PECOS RIVER

Doan's Crossing

Matados

6666

2M2 2M2

Fort Griffin

SHAWNEE

Paris — early home and burial site of John Chisum

Fort Worth

Dallas

By 1900 an estimated 800,000,000 prairie dogs destroyed enough grass to feed over 3,000,000 cattle, which triggered eradication efforts.

El Paso

GOODNIGHT-LOVING TRAIL

During the 1850s Texas longhorns were driven to the California gold fields.

Fort Concho

Horsehead Crossing

Waco

Lampasas
Horrell-Higgins Range War

1870 — Toll bridge opened across Brazos River to accommodate trail herds. Built by Roebling brothers (who later built the Brooklyn Bridge), the 475' span was world's longest suspension bridge.

Before the Civil War Texas longhorns were driven to New Orleans →

Austin

Bandera

San Antonio

NUECES RIVER

During the 1760s Mission La Bahia herded 15,000 cattle and Mission Rosario 10,000

Rosario
La Bahia

Rancho Grande

Houston

Galveston

GULF OF MEXICO

By the end of the Civil War 5,000,000 longhorns roamed South and West Texas.

Although Texas was penetrated by railroads during the 1870s, trail drives continued — it was more economical than shipping by rail, and cattle could gain weight on a slow drive with good grass.

During the 1700s one Spanish rancho along the Rio Grande maintained 750 people and 9,000 cattle and horses.

King Ranch

RIO GRANDE

The Nueces River, Rio Grande and Gulf of Mexico formed a natural corral in an unpopulated area that proved to be a fertile breeding ground for longhorns.

Texas

ᴡ THE KING RANCH

The fabled King Ranch offers more than the story of the growth of an immense cattle kingdom. In many respects, Richard King's magnificent domain provided the creative impulse for the rapid foundation of the cattle industry of the West.

King arrived in Texas in 1847 at the age of twenty-two as a riverboat pilot for Zachary Taylor's army. Following the Mexican War, King, in partnership with Mifflin Kenedy and Charles Stillman, built a profitable steamboat business along the Rio Grande. But his life changed course in 1852, when he rode horseback across the Wild Horse Desert, also known as *El Desierto de los Muertos* (The Desert of the Dead). This harem for mustangs occupied a lonely, drought-stricken expanse of grassland between the Rio Grande and the Nueces River. King immediately recognized the possibilities of raising cattle in the area, and he wasted little time in securing title to more than 200,000 acres of land stretching twelve miles along both sides of Santa Gertrudis Creek.

The Santa Gertrudis was forty-five miles below Corpus Christi in a region long ravaged by hostile Indians and outlaws from Texas and Mexico. Ranchers had been driven out of business by these belligerent elements, and their abandoned cattle had been driven away

The fortified headquarters building of Richard King was erected in 1858, then enlarged in 1868 and in 1909. Today it serves as Wildlife Management Head-quarters of the King Ranch.

—Courtesy King Ranch

by "Cow Boys," young Texans who scoured the Wild Horse Desert for livestock. Far from being intimidated by the failure of his predecessors, King eagerly snapped up land at bargain prices. His first 15,500 acres was acquired for less than two cents per acre. Cattle could be obtained cheaply south of the border. In 1853 and 1854 he began to stock his range with Mexican longhorns purchased for five to seven dollars a head.

A blockhouse and stockade were erected with the first corrals, and cannon from surplus Mexican War steamboats were installed. King employed — for twenty-five pesos per month — hard-riding *vaqueros* who brought with them from Mexico the skills of handling

VOCABULARIO

In the *charro* culture of northern Mexico, a *rancho* was an estate devoted to cattle raising. These longhorn *vacas* were herded by *vaqueros*, a word which was altered by Texans to "buckaroos." In the same way *mesteno* became "mustang," *jaquima* became "hackamore," and *juzgado* (jail) became "hoosegow." The *vaquero* wore a *sombrero* and *chaparreras*, shortened to "chaps." Other *charro* terms which became part of the western vocabulary included *bronco, corral, la reata, lazo, rodear, remuda, caballado, adobe,* and *arroyo.*

cattle from horseback. Having cleaned out one poor Mexican village of cattle, King invited the inhabitants to migrate to his ranch. These people formed the nucleus of the *Kiñenos* who would provide the King Ranch with generation after generation of loyal employees.

King's riders in the early years had to be handy with a gun, and Capt. James Richardson, Mexican War veteran, was hired to ramrod this tough crew. Another early associate was Capt. G. K. "Legs" Lewis, a partner of King's who was removed from the picture in 1855, slain by a cuckolded husband.

A more lasting partner proved to be Henrietta Chamberlain, daughter of a Presbyterian minister in Brownsville. The black-bearded King, a profane and rough-mannered riverman much given to fisticuffs, courted the preacher's daughter for four years before she consented to marry him in 1854. They drove to the ranch in a specially purchased coach, escorted by heavily armed outriders, to the flimsy *jacal* which would be their home. Henrietta became La Madama, the beloved *patrona* of the great ranch. They maintained a house in Brownsville, where King continued to participate in profitable steamboating ventures, but they were increasingly drawn to the raw but promising property along the Santa Gertrudis. (Of course, the 124-mile journey from Brownsville was hazardous. Once Henrietta King was confronted in the *jacal* by an Indian, who threatened her infant daughter with a war club until the alarmed mother gave him all the food he could carry.)

A frequent visitor in Brownsville and at the ranch was Lt. Col. Robert E. Lee, who became friendly with the Kings while stationed in Texas. Lee, a veteran of the Corps of Engineers and an expert judge of terrain, selected the site of the "original" ranch house, where the spacious headquarters now stands. Erected in the late 1850s, the so-called "original" house was a rambling story-and-a-half frame structure, boasting a long, banistered front gallery and a separate stone kitchen and dining room in the rear. Numerous outbuildings included dormitories, sheds, shops, corrals, a watchtower above a large brick commissary, several brick cisterns, and, by the late 1860s, a one-room schoolhouse for the ranch children. Nearby King had directed the damming of Tranquitas Creek when he first began to organize his ranch, and in this vicinity the earliest versions of roundups were conducted. King used a number of brands —

such as *K* (Lewis and King) and *HK* (Henrietta King) — but by 1869 he had adopted the famous Running W.

By the outbreak of the Civil War, King had accumulated 20,000 cattle, flocks of sheep and goats, swine, mules, and 3,000 horses (he loved fine racehorses, and a large remuda of working horses was a necessity for a cattle operation of his size). Early in the war King longhorns apparently were driven to New Orleans, and throughout the conflict military needs provided a market for horses. As the Union tightened its blockade of the Confederacy, southern cotton was funneled to faraway Brownsville, and the heavily traveled cotton road passed within sight of King Ranch headquarters. The ranch supplied the cotton caravans with horses, mules, and beef and other provisions for the final lap of their journey. King bought some of the cotton himself on speculation, and his steamboat partnership continued to be immensely profitable.

Late in 1863 a Union invasion struck the ranch headquarters and killed a faithful *vaquero,* but King was absent. Despite the presence of Union forces at the mouth of the Rio Grande, King continued to make money by running his steamboats under a Mexican flag. During the war King and his riverboat and ranching partner, Mifflin Kenedy, acquired still more land for their ranch.

King and Kenedy soon decided that an amicable division of their property was desirable before it became too vast and inextricable, and in 1868 they dissolved their eight-year partnership. Kenedy assumed control of the Laureles Ranch, twenty-two miles below Corpus Christi, while King was free to develop the Santa Gertrudis property into what would become the most fabled ranch in the world. Scorning the open range system, King and Kenedy began to fence their neighboring ranges, using creosote posts with three pine planks between them. When Kenedy completed a three-plank fence across a peninsula of the Laureles, enclosing 131,000 acres, he became the owner of the first fenced range of real size west of the Mississippi River. Roundups and division of King and Kenedy livestock took most of 1869, by which time King had decided to turn his back on the declining river trade and concentrate his considerable energies and talents upon ranching.

During the post-Civil War decade, King's greatest problem was posed by rampant cattle theft. With the Texas Rangers disbanded by Reconstruction decree and federal military protection at a minimum, border rustlers in uncontrolled numbers drove cattle from South Texas and sold them in Mexico. Hide-peelers skinned cattle — sometimes while the poor beasts were still alive — on their home ranges and made off with valuable hides. From 1866 to 1869, King lost more than $2 million worth of property, and during the next three years he reported the theft of 33,827 head of cattle. Many ranchers were burned or driven out during this lawless rampage, but King fought back ferociously.

The lookouts were redoubled atop the watchtower above the ranch commissary, and the brass cannon were kept ready to fire. King bought thirty stands of Henry rifles and hired more gunmen to ride patrol and pursuit missions under James Richardson. At least once, as late as 1875, ranch headquarters was besieged by outlaws, and ambushes were so frequently set for King that he had to travel with an armed escort of at least a dozen men; indeed, in 1872 a passenger sitting beside King in his coach was shot to death during a fusillade. (King sometimes carried as much as $50,000 cash in a secret box built into his coach.) For the long, hazardous trip to Brownsville, King maintained relay stations about twenty miles apart to provide speedy remounts and fresh teams.

King stepped up his fencing program, and he and Kenedy became the founders of the Stock Raisers Association of Western Texas, which attempted numerous countermeasures against the marauders. The violence was not quelled until the mid-1870s, but by 1869 King had begun to send large herds of longhorns north to the Kansas railheads. His "Kansas Men" (King's terminology), led by a "herd boss" (who was given profit-sharing opportunities), drove the tough longhorns up the "road" (not "trail"), while King traveled ahead to the marketplace to negotiate the sales personally. He raised further operating funds through the sale of horses and mules, wool, and the operation of a hide and tallow works, to which culls were taken as King upgraded his cattle herds. By this time King employed more than a hundred riders, who stayed busy branding 15,000 calves annually and tending a total of 50,000 cattle, 30,000 sheep, and perhaps 6,000 hogs.

The Kings had three daughters and two sons, and each year, usually in coordination with a business trip of her husband, Henrietta took the family on an extended vacation. They often headquartered at St. Louis, but one memorable trip took the family to

12 BRIDES + 12 GROOMS = 24,000 ACRES

Panchito Gonzales of Duval County, Texas, owned a 24,000-acre ranch that had originated as a Spanish land grant. When his oldest child decided to marry early in the twentieth century, Panchito staged a memorable wedding *fiesta*, then gave the young couple title to 1,000 acres. The following year there was another Gonzales marriage, another *fiesta*, and another 1,000-acre wedding gift.

Panchito and his wife were blessed with twenty-four children, twelve boys and twelve girls, and the Gonzales wedding *fiesta* became an eagerly anticipated annual event in Duval County. The 1,000-acre gift to the newlyweds also became traditional, and when the final wedding took place, Panchito retained only three acres around the old home place. There he lived in retirement, surrounded by two dozen children and hordes of grandchildren. (Related by former King Ranch cowboy Emeterio Barrera, nephew of Panchito Gonzales.)

Kentucky, where Captain King bought blooded stallions for the ranch and where the Kings visited an old friend, the president of Washington College at Lexington, Robert E. Lee. (The youngest King child, then four years old, was named Robert E. Lee King.) When they reached their mid-teens, the children were sent to school in St. Louis.

In 1882, troubled by drought, health problems, and by the recent death of his most promising successor, nineteen-year-old Robert E. Lee King, Captain King put the vast Santa Gertrudis Ranch up for sale. A deal was struck with a British syndicate, but they could not raise King's asking price of $6.5 million. He never again considered selling his magnificent ranch. Employing lawyers with a scrupulous eye toward clear titles, King continued to purchase land, finally amassing more than 614,000 acres.

But by the time he was sixty, Richard King was wracked by stomach cancer. He died in San Antonio's Menger Hotel in 1885, willing his ranch and miscellaneous business interests to his wife, Henrietta, who also inherited a debt of approximately $500,000. She appointed as ranch manager Robert Justus Kleberg, King's most trusted lawyer and the fiancé of Alice Gertrudis King. Alice and Kleberg married the next year, and eventually had five children. Mrs. King accompanied the newlyweds on their honeymoon.

La Madama survived her husband by four decades, and even though she backed Kleberg's progressive policies to the hilt, she was the ultimate source of authority until her death in 1925 at the age of ninety-two.

After the ranch house was expanded following the Klebergs' marriage, ten bedrooms were added, in part to accommodate a constant stream of visitors, and the dining table sat twenty-eight. Mrs. King, clad in black for the rest of her life, invariably sat at the head of the table, with Kleberg at the other end and guests and family members in between. In 1893 Mrs. King built an ornate Victorian house in Corpus Christi, where her grandchildren lived during their school days.

Shortly after assuming management of the King Ranch, Kleberg began improving the vast pastures. He cleared the range of mustangs, once gathering 4,000 of the scrubby ponies and trading them for a few head of quality mounts. By the turn of the century the ranch was one of the world's largest producers of horses and mules, selling thousands of horses to police departments and to the United States and Mexican armies, and supplying thousands of farmers with plow mules. Kleberg cross-fenced the range with barbed wire, dividing the ranch into closures with shade and a water hole or well. He began drilling activities in 1899; in time there were seventy-five artesian wells and 225 windmills in operation. In 1891 Kleberg built the first cattle-dipping vat and developed a wash which proved instrumental in eventually ridding western ranges of "tick fever." He also discovered that buzzards were primarily responsible for spreading anthrax.

James Doughty, the longtime chief foreman, died in 1892, and Kleberg hired Sam Ragland, an expert cattleman and bachelor who devoted the rest of his life to the King Ranch. Kleberg also employed cousin Caesar Kleberg, who in nearly half a century of service proved to be a key figure in the ranch organization. Another legendary employee was Faustino Villa, who had worked as a deck hand under Captain King before entering the ranks of the original *Kiñenos.* At the age of 100, Villa swam half a mile across the flood-swollen Santa Gertrudis Creek, and he rode horseback until a fortnight before his death in 1929 at approximately 118.

The ranch continued to purchase large amounts of acreage, with the usual careful attention to titles. In 1907 an investment group, motivated in part by recent oil discoveries in Texas, offered $10 million for the ranch, but Mrs. King declined the overture. By that time, 75,000 cattle grazed on 1,150,000 acres, along with 10,000 horses and mules. Ranch work was divided among the Santa Gertrudis, Laureles, and Norias divisions.

During the early 1900s, after a railroad finally was built through the area, the King Ranch organized and promoted the town of Kingsville alongside the tracks. The five Kleberg children continued to attend school in Corpus Christi; each weekend an engine and caboose would take them back to Kingsville, where horses were waiting for a ride to the ranch, and at the first of the week they would be whisked back to school aboard their tiny private train. In 1912 the sprawling frame house was destroyed by fire, to be replaced by a twenty-five-room mansion of Spanish-Moorish motif which cost $350,000. Three years later a somewhat different conflagration erupted at Caesar Kleberg's two-story headquarters of the Norias Division. Nearly sixty Mexican bandits were held at bay by a handful of defenders during a furious rifle battle.

After the death of Henrietta King in 1925, the twenty-two-page will stipulated a ten-year period of trusteeship prior to partition of the estate by the heirs. A precise inventory tallied more than 94,000 head of cattle grazing upon about 1,150,000 acres of land, and during this period about 25,000 calves were branded annually. When the partition finally was worked out, the King Ranch, under management of Bob Kleberg (his father, Robert, had died on the ranch in 1932), retained 60,000 head of cattle on 890,000 acres. There also are thousands of acres of lush pastureland owned in Pennsylvania and Kentucky, along with millions of acres in Australia and Latin America.

Chief among twentieth-century achievements on the King Ranch was the development of the first distinctive American breed

After the sprawling frame ranch house burned in 1912, a twenty-five-room mansion was built at a cost of $350,000.

—Courtesy King Ranch

Carriage house and stable, built in 1909.

—Courtesy King Ranch

Lolo Trevino, a Kiñeno *for half a century, mounted on Camote ("Sweet Potato").*

—Courtesy King Ranch

of cattle, a mixture of three-eighths Brahman and five-eighths Shorthorn. The foundation sire was a remarkable red bull, named Monkey. He was born in 1920 and produced over 150 useful sons. The breed he sired boasts high quality beef characteristic of Shorthorns, plus the loose, tough skin of Brahmans which resists pests and heat. Also from Brahman blood is the ability of the Santa Gertrudis to gain weight quickly on grass, as well as survivability qualities in harsh climates.

Another breeding triumph was the development of the King Ranch quarter horse. The forebear was Old Sorrel, a stud colt acquired in 1916 by Caesar Kleberg. King Ranch quarter horses are in great demand for cow work, polo ponies, superb saddle horses, or, of course, champion entrants in quarter-mile stakes. Captain King brought the first thoroughbreds to the ranch, and Bob Kleberg entered thoroughbred racing with such enthusiasm and scale that the King Ranch stable has won virtually every prize, including the Triple Crown. The King Ranch also was noted for a special training method: horses are never "broken" but gentled. Instead of employing bronc busters, the King Ranch historically has fostered a policy of gradually accustoming horses to handling from the time they are foals. By this procedure horses are trained rather then subdued.

A significant twentieth-century occurrence was the discovery of oil by Humble, now Exxon. Nearly 400 wells pump oil from beneath the ranch, but the proceeds always have been channeled back into range improvements and superior livestock.

Heading up the ranch throughout most of the twentieth century (1918-1974) was Bob Kleberg, energetic and creative grandson of the founder of the famous old spread. Under his dynamic leadership and continuing today, the King Ranch has maintained its preeminence in cattle ranching, constantly researching and experimenting in genetics, grasses, mineral feeding, watering, fencing, wildlife, and conservation. The home range was uninviting and hardscrabble in many respects, but through the classic pioneer qualities of hard work, perseverance and innovativeness the Kings, Klebergs, and *Kiñenos* established an institution that for more than 140 years has blazed the trail of the West's most colorful industry.

𝒜   THE JAS

The founder of the JA Ranch, a spread always referred to as "the JAs," was legendary frontiersman Charles Goodnight. In 1845, at the age of nine, he had ridden bareback 800 miles when his family migrated from Illinois to Texas. He later became a jockey, teamster, rail-splitter, Texas Ranger, an experienced scout against Indians and Mexicans, and the most renowned trailblazer of the cattle frontier.

Goodnight opened five trails after the Civil War, the most famous of which was the Goodnight-Loving Trail. He had an uncanny sense of direction and terrain, and his masterful organization and careful precautions earned him safe passage where others encountered various disasters. For six years he operated a ranch near Pueblo, Colorado, where he engaged in multiple business enterprises. But the Panic of 1873 gutted him financially, and within a couple of years Goodnight found it necessary to relocate.

Late in 1875 Goodnight and a *vaquero* called Pancho rode into the vast, trackless Texas Panhandle, which had just been cleared of enormous buffalo herds and hostile Indians. After weeks in the saddle Goodnight located the grassy, well-watered Palo Duro Canyon, whose isolated reaches had been the final Comanche sanctuary, inviolable until Ranald Mackenzie's Fourth Cavalry clambered down into the canyon in September 1874. The sprawling plains above the canyon would afford excellent summer grazing, while the

Palo Duro would provide sheltered winter pasturage. The next year Goodnight eagerly brought 1,600 longhorns to the enormous canyon. He and two companions descended the precipitous Indian trail four miles to the canyon floor, where he was astounded to discover more than 10,000 buffalo survivors

Rancher Charles Goodnight.
—Courtesy Panhandle-Plains Museum, Canyon

Home Charles Goodnight built for his mother.
— Courtesy Panhandle-Plains Museum, Canyon

grazing. The three riders drove the buffaloes fifteen miles down the canyon, keeping the roaring herd on the move with constant gunfire. As a cloud of red dust rose 1,000 feet to the rim of the caprock surrounding Palo Duro, the canyon walls swarmed with black bears, panthers, deer, turkeys, ducks, and other disturbed wildlife. Goodnight and his cowboys then spent two days packing their dismantled wagons and six months' supplies down the dangerous trail by mules, after which the longhorns were driven single-file to the core area of the first real cattle ranch in the Panhandle.

Goodnight needed financial backing to expand his ranch on a grandiose scale. In Denver he was introduced to John Adair, a British financier who was willing to add to the flood of English money being pumped into western ranching. In 1877 Adair and Goodnight entered a five-year agreement: Adair would furnish financing, while Goodnight would provide the foundation herd and manage the ranch at $2,500 per year; at the end of five years the assets would be divided, one-third to Goodnight and the remainder to Adair; and Goodnight agreed to repay Adair's investments in cattle and land at ten percent interest. Questioned later about the lopsided agreement — particularly the high interest rate — Goodnight revealed: "I did not mind it, because I knew I had a fortune made."

Settlers were rapidly migrating to the Panhandle, but Goodnight began a masterful series of "crazy quilt" land purchases. Following careful surveys of sections with good water and grass,

JAs cowboys.
— Courtesy Panhandle-Plains Museum, Canyon

Goodnight bought choice sites for seventy-five cents an acre. Free grazing was available on every other section (alternate sections in the area had been set aside for Texas schools), and the JA monopoly of these sections, combined with the pattern he already had purchased, allowed Goodnight to snap up the remaining lands for twenty to thirty-five cents an acre. Goodnight bought the Quitaque Ranch (and a cattle herd branded with Lazy F's), a fine range adjacent to the south, for twenty-two cents an acre, and he purchased the Tule Ranch in 1883. By the mid-1880s more than 100,000 JA cattle grazed on more than 1,335,000 acres.

An early believer in barbed wire, Goodnight carefully fenced the JAs. He suggested the 𝒥𝒜 brand because of Adair's initials, although the ranch's finest cattle wore the 𝒥𝒥 brand. Goodnight placed his pure-bred livestock in the upper range of the Palo Duro and branded them with the 𝒥𝒥. The best bulls were put with the 𝒥𝒥 herd, and any inferior offspring joined the main herd. Cattle purchased outside the JAs were carefully culled: serviceable cows went to the main herd while inferior stock were cut out and spayed, fattened on JA range, then trailed to market in the fall with the beef herds. Goodnight thus doubled the speed of improvement of his

Originally a water tank built on the JAs to water 3,000 cattle, this structure was converted to an oat bin in the 1920s. Now it is part of the Ranch Heritage Museum in Lubbock.

herds while constantly upgrading their quality.

Goodnight was a forceful disciplinarian, imposing three rules upon his riders: no drinking, no gambling, no fighting. The steady JA hands saved their wages (in 1885 there was $26,000 on deposit at the ranch for employees), and the cowboys were allowed to run their own horses and cattle on JA range.

Goodnight's first quarters in Palo Duro was a dugout, but soon he was able to place his wife in a timber structure which would be known as the Home Ranch. The Goodnights were childless, but "Aunt Molly" mothered her husband's riders, and Goodnight later gave her a tall clock with a lengthy inscription praising her for cheerfulness and courage in standing up to hardships and complete isolation from other women. Goodnight invented the first safe sidesaddle for his wife, and also developed the first chuckwagon. In 1879 headquarters was moved twenty-five miles to the east, and a number of commodious buildings were erected closer to civilization.

Goodnight secured his range by reaching an agreement with sheepherders in the Canadian River area to stay out of their pastures if they would avoid the Palo Duro vicinity. Rustler Dutch Henry Born presented another problem which Goodnight met with characteristic directness. He parleyed with the outlaw leader, informing Dutch Henry that he commanded a large number of well-armed riders but that he preferred not "to use them in that way" and would not if Born's men would stay away from his range. Goodnight and Born sealed their pact with a drink, and Dutch Henry's gang caused no trouble on the JAs. In 1880 Goodnight spearheaded the organization of the Panhandle Stock Association of Texas, as a means of applying widespread pressure against rustlers. The next year, in order to halt downstate herds headed north from spreading tick fever as they passed through the

The main house at the JAs was built for John and Cordelia Adair, and still is used by current owners.

The post office at the JAs, where mail is delivered three times a week.

Boot scraper in front of the day room at the JAs.

House at the JAs built for Charles and Molly Goodnight.

Panhandle, Goodnight enforced the "Winchester Quarantine" with heavily armed riders posted to turn back cattle from the south.

Goodnight's leadership produced impressive results. In 1882, at the end of the five-year Adair-Goodnight contract and after Adair was repaid according to agreement, there was a profit of more than $512,000. Adair pressed for a new five-year pact: ranch manager Goodnight would be paid $7,500 annually, and Adair would be repaid eight percent interest. Goodnight agreed to these terms, but Adair died in 1885. By this time settlers and politicians were erecting strong pressures against large-scale ranching, and Goodnight, beset with acute abdominal problems, wrote Adair's widow in London that he wanted to divide the JAs.

Impatient to dissolve the partnership, Goodnight offered Mrs. Adair the superb Palo Duro rangeland, a decision he later regretted deeply. During more than a decade on the JAs, Goodnight had built nearly fifty ranch houses, bunkhouses and line cabins, hundreds of miles of roads and fencing, two dozen stock tanks, a large blacksmith shop and tin shop, and he had developed a dairy, hay farm, and poultry yard. Goodnight settled for the 140,000-acre Quitaque Ranch, which he soon sold. At the end of 1887 he left the JAs with his wife for Goodnight, a little Panhandle town on the railroad named in his honor by the Fort Worth & Denver. The Goodnights built a rambling frame ranch house there which still stands and where the famed cattleman spent the last forty years of his life.

Goodnight brought a buffalo herd from the JAs to his new home place. In 1878 he had begun preserving the rapidly disappearing animals, and his bison herd eventually grew to 250. The most famous animal on the JAs, however, was Old Blue, a 1,400-pound longhorn steer which was used again and again by Goodnight as a trail herd leader. When a trail drive commenced, a brass bell was attached to Old Blue's neck. The herd animals became so accustomed to following the bell and the tireless strides of Old Blue that at night the bell was muffled, since the sound of the clapper would bring the herd to its feet, ready to travel. Old Blue could lead a herd thirty miles in a day, he never participated in stampedes, and sometimes he was instrumental in bringing an unruly herd under control. The steer was shipped back to the JAs after each drive, and eventually he was turned out to pasture for an honorable retirement. Another noted JA animal was Old Maude, a Texas cow which was a

JAs horse in front of the feed room at the stables.

member of Goodnight's original herd in the Palo Duro. Old Maude had twenty-seven calves and, like Old Blue, also was given retirement range. When Old Maude lost her teeth, one of the women on the ranch fed her mush.

After Goodnight left the JAs the ranch settled into a more prosaic existence. Adair heirs maintained the ranch, still running 25,000-30,000 cattle on nearly 500,000 acres through the middle of the twentieth century and 14,000 to 20,000 head on 397,800 acres in later years.

El Rancho Grande

"I am Shanghai Pierce, Webster on cattle, by God, sir."

The big, flamboyant rancher who boomed out this announcement to a hotel clerk was, indeed, cattleman enough to become legendary throughout the West. He would receive credit for introducing the Brahman breed to Texas in order to combat the deadly "tick fever." Shanghai organized El Rancho Grande in 1865, and when he died thirty-five years later his first big ranch still was in operation.

His younger brother, Jonathan, was in charge of the ranch at the time of Shanghai's death.

Abel Head Pierce was born on a Rhode Island farm in 1834; Jonathan Edwards Pierce came along five years later. The Pierce ancestry went back to the Mayflower, but Abel chafed under a strict New England upbringing. He went to work for an uncle, Abel Head, who was a merchant in Virginia, but soon stowed away on a schooner bound for Texas. Although the six-foot-four-inch stowaway was quickly discovered, he earned his passage to Indianola as a cargo handler. The loud nineteen-year-old then found employment as a rail-splitter on a Matagorda County ranch, asking his boss for his pay in cows and calves because, "I want to be a cattleman like you some day."

Acquiring the sobriquet "Shanghai" from obscure sources, the aggressive youth became an impressive bronc buster and learned the cow-

Shanghai Pierce commissioned San Antonio sculptor Frank Teich to produce a gravesite memorial "higher than any statue of any Confederate general." Pierce paid Teich $2,250, but liked to brag that the statue cost $20,000.

boy's skills. He also learned how to build a herd by the "grab game," grabbing mavericks on the open range and applying the AP brand from an iron he had made. Continuing to master the cattleman's trade, Shanghai bossed a trail drive to New Orleans before the Civil War. During the war the ex-Yankee spent a couple of years in a Matagorda County military unit, assigned (because of his known expertise as a rustler) to gather cattle for Confederate commissaries. When he teased about his service as a regimental butcher, Shanghai replied: "By God, sir; I was all the same as a major general: always in the rear on advance, always in the lead on retreat."

Jonathan Pierce had joined his older brother in Texas in 1860, and he spotted a site beside Tres Palacios River in western Matagorda County where he wanted to make his home. Shanghai, brimful of confidence in his abilities ("I stand pat I am the best cowman in Texas today"), proposed a ranching partnership to his brother. Jonathan would own one-fourth of the livestock, while Shanghai would own three-fourths and do all of the cattle buying and selling and other outside work. "It is understood, John, you are to stay at Rancho Grande all the time."

The brothers had letterheads printed in Galveston and otherwise advertised their grandiosely named ranch. With relentless ambition Shanghai rode his big mount, Old Prince, in constant search of deals. Although he wed a neighborhood widow, Fannie Lacy, on September 27, 1865, he later boasted of his work habits (if not his marital devotion): "When I married I only stayed home four days the first year."

Jonathan married Nannie Lacy, sister of Shanghai's wife, on May 2, 1866. Jonathan announced to Shanghai, "I have just become your brother-in-law." The couple eventually would have four children. Jonathan built their home himself at his dreamsite, pointing out "there's 2,200 pounds of galvanized nails in it, and the walls would stand a siege."

Soon a frame office was constructed, then a barn, a saddle house, two carriage houses, a store, a smokehouse, and two bunkhouses, one for whites and the other for black cowboys. There would be cattle pens, a corn crib, a potato house, a hog house, a goat house, a carpentry shop, a hide press, a hide house, and a blacksmith shop.

The foreman at El Rancho Grande was tough, hard-drinking, pipe-smoking Wiley Kuykendall, who had married Susan Pierce after she joined her brothers in Texas. His crew was made up of war-hardened Confederate veterans, Mexican *vaqueros*, and a large number of black cowboys. There were also teenagers in love with horses and range life, such as sixteen-year-old Charley Siringo, who signed on in 1871.

For thirty-five years Shanghai was faithfully attended by a black servant, Neptune Holmes. Riding a mule, Old Nep accompanied Shanghai on his constant travels. Shang always carried a Winchester, as well as a lariat, in case he spotted a maverick to brand.

Soon the Pierce brothers were assessed taxes on 35,000 branded cattle in Matagorda County, and in one year El Rancho Grande branded 18,000 calves. Shanghai also entered partnerships with sev-

A ROSE
IS A ROSE IS A . . . FENCE

In 1867 the wild rose was introduced to Matagorda County, and Jonathan immediately began planting the bushes as an enclosure around El Rancho Grande. Jonathan reasoned that the plants would keep out rustlers, because any "thief who will penetrate a rosebush hedge is hardly fit for business after the trip."

As the plants flourished, thick hedges were formed which outlined the various pastures and which created windbreaks. During harsh weather the cattle sought shelter among the hedges.

Another benefit was described by Jonathan: "If you want to see the earth in all its glory, come here in the spring. The whole thirty-seven miles of hedge is then one great blaze of rose blossoms and the perfume of the air hereabouts is so sweet that those who visit me say they are in paradise."

eral other cattlemen, most notably Allen, Poole and Company. Later Shanghai would boast, "I owned nearly all the cattle in Christendom once."

But so many cattle inevitably attracted rustlers and brand burners. Trouble came to a head in 1871, as Shanghai issued pointed warnings, then received death threats. When a Pierce cowboy rode into headquarters at El Rancho Grande with news that five thieves were lurking in Newell's Grove with hides from stolen cattle, Shanghai sent riders to his camps and to other ranches. Before dawn nearly eighty men closed in on Newell's Grove, capturing five thieves, then hanging them from a dead tree. "Well, sir," remarked Shanghai, "you never can tell just how much human fruit that old dead tree might have borne had it only been green."

Although the identity of the hangmen could not be established, legal action was commenced against Shanghai, and he was placed under a $200 bond. He quickly sold his partnership to Allen and Poole for $110,000 in gold and credited Jonathan with his personal indebtedness for $10,000, giving the younger brother control of El Rancho Grande. With $100,000 in his pockets, Shanghai left Texas for Kansas City, Omaha, and Abilene.

Shanghai was gone for a year and a half, buying and trading cattle, before Jonathan wrote him that "atmospheric conditions" in Matagorda County had cleared enough for Shanghai to return. The older brother finally went back to El Rancho Grande.

Mamie Pierce was born to Shanghai and Fannie in 1867, but shortly after giving birth to a son in 1871, both mother and baby died. Jonathan and Nannie cared for "Miss Mamie," before Shanghai sent her to stay with one of his sisters back East. When Shanghai returned to Texas he sent for Miss Mamie, then soon remarried.

His new wife was Hattie James, who was in charge of the property of her insane father. Following a honeymoon in Galveston, Shanghai determined to put together another ranch in Wharton County, to the northeast of El Rancho Grande. He formed a partnership with Daniel Sullivan and became president of the Pierce-Sullivan Pasture & Cattle Co. Shanghai pursued the project with his customary frenzied energy, building his new home a few miles south of Wharton. Purchasing tracts totaling 250,000 acres, he annually sent thousands of head of cattle north to market.

Jonathan remained in charge at El Rancho Grande, although he continued a partnership arrangement with Shanghai on various cattle deals. Occasionally Shanghai would travel to a northern market and contract to sell a certain number of cattle, then would let Jonathan know how many cattle to ship within prescribed price limits. If Jonathan could not gather enough cattle at El Rancho Grande and at Shanghai's ranch, he resourcefully purchased livestock from neighboring spreads.

Wiley Kuykendall stayed at El Rancho Grande as foreman, and Jonathan put some of his land under cultivation while operating the ranch profitably. Determined to find a breed that would withstand "tick fever" or "Texas fever" (Texans called it "Spanish fever"), Shanghai finally settled upon India Brahmans, although the first importations of these superb animals were not effected until after his death.

Shanghai died of a cerebral hemorrhage in 1900 at his Wharton County ranch. He was sixty-six. Jonathan's wife died, too, but he twice remarried and had two more children.

When a railroad crossed El Rancho Grande just south of head-quarters, Jonathan exclaimed, "Thank God!" He wanted to name the new townsite southwest of headquarters Thank God, but was persuaded to call it Blessing. Jonathan lived until 1915, dying at the age of seventy-five on his beloved El Rancho Grande. Among other blessings, oil was discovered on the old ranch, and today black gold still pumps from the former Pierce pastures.

ᴸS LEE-SCOTT CATTLE COMPANY

W.M.D. Lee and Albert Reynolds, who had made a fortune as partners in the freighting business, founded the LE Ranch in the 1870s when Texas Panhandle grasslands were cleared of buffalo herds and hostile Indians. The partners soon quarreled and Reynolds bought out Lee, who determined to stay in the cattle business in the Panhandle. Lee found a new partner, Leavenworth businessman Lucien B. Scott, and the Lee-Scott Cattle Company soon was established, destined to scatter livestock over rangeland nearly the size of Connecticut.

Lee and Scott concluded that free grass soon would be a thing of the past, so they began to purchase land for their ranch, securing title to 221,000 acres in Oldham, Potter, and Hartley counties by the time full-scale operations commenced. For years a number of sheep-men from New Mexico had grazed their flocks in the vicinity, erect-ing residences and pens which became known as *plazas.* Sharing the cattleman's typical distaste for sheep but scorning the use of force, Lee stuffed $35,000 in currency in a valise, hitched a team to a buggy, and drove alone into *plaza* country, heedless of danger. At each *plaza* Lee offered cash for removal to New Mexico, and by 1884 the sheep-men had retired peaceably. LS cattle soon were grazing across Oldham and surrounding counties and well into New Mexico. An early LS cowboy stated, "No matter where you begin on this ranch, it's a long ride before you start coming back."

To stock their spread, Lee and Scott rapidly bought and con-tracted for large herds of cattle, from four- or five-year-old long-horns to a herd of Durham bulls Lee purchased in Scotland. Foreman J. E. McAllister, who had been Lee's ranch manager on the

LS headquarters was relocated in the 1890s to a new complex four miles south of Tascosa.

Second LS headquarters, twelve miles south of Tascosa. J.E. McAllister and his bride lived in the residence on the left, while the bunkhouse is on the right. This layout was suggested by an Arbuckle coffee label.

Branding time on the LS. Two stamp irons were used, one bearing an "L" and the other an "S."

LE, had to assemble an outfit to handle the influx of livestock. McAllister gathered six chuckwagons and assigned a boss, cook, ten cowboys, and a wrangler to each wagon. During that first year not a corral or a branding pen had yet been erected, and each outfit had to work incoming herds in the open. Several riders bunched and held the cattle while others branded (two men simultaneously applied stamp irons bearing an "L" and an "S"), earmarked, and tallied each animal before turning it out to pasture. By the end of the year 50,000 cattle grazed on LS range. During its formative period the LS employed 150 riders, with common cowboys earning $40 a month beginning in 1886. (One young cowboy was Ed Doheny, who later became an oil magnate and who was involved in the Teapot Dome scandal during the Harding administration.) There were eventually ten chuckwagons: four for trail drives, four "floaters" which constantly worked one area of the LS after another, and two "extras."

THE COWBOY STRIKE AND THE HOME RANGERS

LS wagon boss Tom Harris led the famous cowboy strike of 1883 in the Tascosa vicinity. Harris made $100 a month, $25 more than the strikers were asking for men who ran an outfit. Lee fired Harris, and the big ranches of the Tascosa area blacklisted the striking cowboys. Hard feelings resulted, aggravating an already serious rustling problem in the vicinity. In 1884 the large-scale ranchers engaged Pat Garrett to lead a company of Home Rangers, which headquartered at the LS and cracked down on rustlers, although the presence of hired gunmen further increased tensions between small operators and the big outfits. The Home Rangers were disbanded in 1885, but most of the riders stayed on to work for the LS. Their continued presence was resented, and on March 20, 1886, a vicious late-night gunfight erupted in Tascosa. Four men were slain, including a trio of LS cowboys who had been Home Rangers. The entire LS crew, spoiling for revenge, came to town as the coffins of their dead comrades were loaded into a wagon and taken to Tascosa's Boot Hill, but McAllister ordered his men to cause no further trouble.

Early in 1884 Lee and Scott, like other Texas cattlemen, organized a "finishing ranch" in Montana. Texas steers in Montana rapidly gained weight and size and brought higher prices from eastern cattle buyers. All LS range in Montana was leased or was public domain. Al Popham was appointed general manager, and headquarters was established thirty miles north of Miles City. That spring two herds of 2,500 steers were trailed to Montana; trail bosses were given letters of credit to resupply en route or take care of unexpected expenses.

In Texas the first headquarters of the LS was on Alamositas Creek, just north of the Canadian River. The structures there were purchased from Capt. Ellsworth Torrey, who reputedly decided to sell his ranch following threats from Billy the Kid and four fellow rustlers. Within a couple of years, however, the establishment of the gigantic XIT Ranch forced a shift of the LS range to the southeast; this development separated the LS from range in eastern New Mexico, and Lee and Scott were forced to sell their holdings there. A new headquarters was built twelve miles south of Tascosa on a stream called Ranch Creek. A pair of two-story stone buildings facing each other — a bunkhouse and the ranch manager's residence — were suggested by Jim East, who had seen a similar design on a sack of Arbuckle coffee, the universal stimulant of western cowboys.

McAllister moved his bride into the residence early in 1886. "Mister Mac" had met "Miss Annie" on a Christmas trip to Iowa in 1884, and in September 1885 he had brought her and a chaperone to the LS for what turned out to be a successful courtship. Mrs. McAllister refused to have the cowboys' minds contaminated by trashy reading material, so she supplied the bunkhouse with subscription copies of the *Christian Herald* and *Leslie's Monthly*.

In 1886 McAllister warned the owners that range was overstocked, but when his advice was ignored he resigned. During the unprecedented blizzard conditions of 1886-87, like most other ranches the LS suffered great losses. McAllister was replaced by Jim East, the former sheriff of Oldham County. Mrs. East, while alone in the residence one night, triggered a shot at "a prowler" — and killed the finest bull on the LS.

In 1888 Lee and Scott sold their Montana spread, and Al Popham was brought to the home range as manager. Lee became interested in other investment schemes, sold his share of the LS to

Scott in 1890, then lost his fortune in an ill-conceived plan to construct a deepwater port and docks at the mouth of the Brazos River.

Scott, determined to avoid the fiasco of 1886-87, embarked upon systematic fence construction, the placement of Wooden Eclipse windmills across his range (at an average depth of 125 feet), and other improvement measures. Popham refused right of transit to other herds, feeling that the LS could not spare grass to be trampled by outside cattle, but the north-bound trail through Tascosa was closed by this step. The economy of the fenced-in trail town was wrecked, and during the 1890s the "Cowboy Capital of the Panhandle" became a ghost town of crumbling adobe buildings.

In 1893 Scott died, and within a few years his widow presented the massive ranch to her brother, Charles Whitman, who had served the LS in various capacities through the years. Whitman built a new ranch headquarters (complete with icehouse and running water) four miles south of Tascosa, persuaded McAllister to resume his old job as ranch manager, and initiated the Christmas custom of giving each LS cowboy a "Whitman" — a fine new Stetson. To combat rustlers, Whitman also purchased a dozen powerful 8mm Mannlicher Austrian Cavalry rifles and ordered his cowboys to shoot strangers on sight. He also maintained a pack of greyhounds to hunt wolves and coyotes. Big lobos could kill grown cattle, and coyotes preyed on calves, but Whitman effectively countered the predators with his seventy-five greyhounds and by offering a ten-dollar bonus to LS riders for every slain wolf. Cowboys enjoyed an assignment as "mush-pot wrangler"; despite having to tend the kennels and mix huge quantities of mush from cornmeal and meat scraps for the greyhounds, the mush-pot wrangler frequently enjoyed the zest of the hunt, and was given a string of the finest mounts on the ranch to ride with the pack.

During the 1890s Whitman, convinced that whitefaced Herefords were the finest beef cattle in the world, culled nondescript cattle from his range and regularly purchased purebred heifers for "Salisbury," a bull which became the talk of Texas cattlemen. In the first year of the LS, Lee and Scott had bought seven carloads of Herefords, and by the turn of the century the quality of LS cattle attracted buyers from as far away as Canada.

Whitman died in Denver of peritonitis in 1899. His widow continued to operate the LS for a time, but she married a prominent

Kansas City businessman, and in 1905 they sold the ranch. The ranch was split; however, in 1907 Col. C. T. Herring bought the LS brand and thirty sections of the eastern range. The pioneering days of the sprawling LS were over, but there would be a substantial vestige of the great ranch throughout the twentieth century.

XIT
XIT (CAPITOL SYNDICATE RANCH)

In 1879, following legislative groundwork laid in previous years, the Texas Legislature passed a law appropriating 3,050,000 acres of Panhandle rangeland to finance a splendid new state capitol in Austin. From this legislation would emerge the Capitol Syndicate Ranch, better known in western ranching circles as the XIT.

The land would extend from the northwest corner of the Texas Panhandle south for more than 200 miles along the New Mexico border, covering parts of ten counties across the sparsely populated Staked Plains. Fifty thousand acres of the sprawling tract were to be sold (at 55½ cents per acre, a good price for Panhandle land), and the Capitol Reservation Lands were surveyed in 1879–80.

Mattheas Schnell of Rock Island, Illinois, was awarded the construction contract in 1882, and within a few months he assigned the contract to a Chicago firm, Taylor, Babcock and Company. This Capitol Syndicate, headed by Abner Taylor, A. C. Babcock, and John V. and Charles B. Farwell, built a magnificent state house in Austin from 1885 to 1888, and as construction work progressed the company received title to their Panhandle lands. The Capitol Syndicate assumed $3,224,593.45 in construction costs, making the price of their land $1.07 per acre, about twice the going rate for well-watered Panhandle rangelands.

A. C. Babcock journeyed to the Panhandle in the spring of 1882 to undertake an inspection and survey on behalf of his company. Employing a crew in Tascosa, Babcock traveled nearly 1,000 miles in thirty-six days and inspected almost all of the designated Capitol Reservation. Babcock returned to Chicago and reported to his partners on soil, grass, and water conditions, along with boundary adjustments which needed to be made. The partners briefly considered selling and colonizing their lands, but soon, carried away with the "Beef Bonanza" fever of that era, decided to stock their range, at least until farmers or other prospective buyers began to move to the area.

To purchase the immense herds necessary for 3,050,000 acres, as well as to provide fencing, houses, barns, corrals, and watering facilities, additional funding was sought abroad. John V. Farwell, who had business connections in Paris, Manchester, and Belfast, traveled to England and in 1885 formed an English company, the Capitol Freehold Land and Investment Company, Limited. Its authorized capital was £3,000,000 (about $15 million), and the investment firm leased the ranch to the Capitol Syndicate, naming John V. Farwell as managing director. The English company had almost nothing to do with operations of the ranch, and in 1909, after the bonds were redeemed, the investment firm was dissolved.

Back in Texas an outfit began to be organized in 1885. Berry Nations was employed as range foreman, Ruck Tanner was the wagon boss, and Englishman Walter S. Maud came over to represent the British investors (although he spent most of his time in Tascosa saloons shooting craps and drinking highballs). The owners secured as general manager Col. B. H. Campbell, generally called "Barbeque" because of his brand, BQ. George Findlay was placed in charge of the ranch's business matters from the Chicago office, and during the next four decades he spent considerable time in Texas.

6,000 MILES OF BARBED WIRE

Fencing operations began in 1884, and within two years 575 miles of barbed wire fence enclosed the XIT. Three hundred carloads of materials were purchased at a cost of $181,000. During the next decade the ranch was divided into ninety-four pastures requiring about 1,500 miles of fence. Some 6,000 miles of wire were used, along with 100,000 cedar posts, five carloads of wire staves, one carload of staples, and an entire carload of hinges for the hundreds of gates. Line riders kept a constant check on the fencing, and some divisions kept fence wagons in operation at all times.

Campbell contracted for numerous herds to be delivered to the ranch during July and August of 1885, and as the herds arrived, a range boss named Collins hired any cowboys willing to stay on, while Tanner purchased any wagons or camp equipment that owners were inclined to sell. Ab Blocker drove the first herd in from the Fort Concho area, and he suggested the brand by which the ranch soon became known. (It is doubtful that Blocker intended XIT to stand for "Ten Counties in Texas"; rather, XIT simply was an easy brand to make and a hard one to alter.)

A corral and a big chute, capable of handling twenty-four cattle, were erected at Buffalo Springs, at the northern point of the ranch. During the first season 22,000 cattle were branded with an XIT on the right side (the famous brand was burned with a five-inch bar, the iron being stamped five times to form "XIT"). An XIT outfit with half a dozen cowboys and a chuckwagon would drive a branded bunch to part of the enormous ranch, loose herd them for several days so that they would "locate" on that range, then return to Buffalo Springs for another bunch.

In the fall of 1885, prairie fires swept across the Buffalo Springs grasslands and other XIT ranges, destroying one million acres of grazing and, along with vicious winter blizzards, costing the ranch more than 5,000 cattle. (Within a year more than 1,000 miles of plowed furrows were placed on both sides of XIT fences as firebreaks; nevertheless, a raging fire in 1887 was stopped only by a snowstorm, and the disastrous blaze of 1894 was the worst of many other prairie fires.) In 1886 nearly 90,000 head of cattle were contracted for, and most of the branding in that year took place at Rita Blanco in Hartley County. The following year was dry, but 30,000 more cattle were delivered to the Yellow House Division (referred to as the Yellow Houses) at the southern portion of the XIT.

Because of unfavorable weather conditions many cattlemen were anxious to sell, and the XIT provided an important outlet, while benefiting from favorable prices during a buyer's market. After 1887 there were no further major purchases of herds by the XIT, although a great many Hereford, Durham, and Polled Angus bulls were brought in from eastern states to improve the beef quality (if not the hardiness) of the rangy Texas longhorns. The large-scale experiments on the XIT with these breeds of cattle made evident the stock-raising advantages of the Panhandle, and by the early

1900s the XIT had developed the largest herd of high-grade Polled Angus cattle in Texas. During its heyday the XIT maintained herds of 125,000–150,000 cattle.

While the long perimeter of the XIT was fenced in by the fall of 1886, well-drilling lagged seriously behind schedule. During the drought year of 1887, as thousands of thirsty cattle were driven to the Yellow Houses, a double row of men passed water buckets from wells to newly built troughs, and the parched cattle fought for position. Wells were dug by hand, and homemade pumps were devised as herds were juggled desperately from one watering place to another. During the 1890s approximately $500,000 was spent on water facilities, including 335 windmills and 100 dams. Earthen water tanks were dug, and several 200-pound sacks of stock salt were spread out along the bottom of each newly completed tank. After cattle and horses had crowded in to eat all of the salt, the tank was well-packed and ready to hold water. Each division employed one or two "windmillers" to drive their wagons from one windmill to another, keeping up with the necessary maintenance.

The XIT erected a 132-foot windmill, tall enough to catch the wind above the canyon walls. This 114-foot replica stands in Littlefield.

Within a few years after the XIT was founded, a shakeup in range management proved necessary. Graft, shoddy managerial practices, and outright theft flourished, while the huge ranch became a known hangout for horse and cattle thieves. In 1887 the Chicago partners ousted Barbeque Campbell and gave A.L. Matlock, a lawyer, state representative, and former district attorney, authority to run the ranch. Matlock engaged A.G. Boyce as the new general manager, and George Findlay came down from Chicago to assist with the crackdown.

Drinking, gambling, and loafing were epidemic among the XIT's fifty riders, but soon Matlock, Boyce and Findlay effected an almost complete turnover in foremen and cowboys. In 1888 twenty-three "General Rules of the XIT Ranch" were posted; among other rules, employees were forbidden to drink, gamble, or carry revolvers. Their conduct was regulated in a variety of ways, although some of the rules were laxly enforced.

General Manager Boyce, assessing the highly varied topographical features of the 200-mile long ranch, in 1887 separated the XIT into seven divisions, numbered from north to south:

(1) *Buffalo Springs* — steers were acclimated for a year on these 470,000 acres before being driven to Montana.
(2) *Middle Water* — a "cut-back" or cull ranch, onto which undesirable cattle were cut out from other divisions and driven to.
(3) *Ojo Bravo* — a breeding range.
(4) *Rita Blanco* — a beef range with excellent grama.
(5) *Escarbada* — a breeding range.
(6) *Spring Lake* — a breeding range where an eighteen-year-old cowboy named Will Rogers worked in 1898.
(7) *Yellow House* — a breeding range.

In addition to the XIT brand, a calf was stamped with the division number on the jaw and the last number of the year on the shoulder (a stray animal with a 7 on its jaw and a 2 on the shoulder had been branded on the Yellow House Division in 1892). Once Boyce had designated the divisions, pastures for different purposes were determined and fenced in. In 1897 land was cut off from Spring Lake Division to form an eighth ranch, the Bovina Division. Each division foreman was given full authority to run his division as a separate ranch, although a monthly report was made to the general manager.

Residences, barns, bunkhouses, storerooms, and corrals were erected at each of the XIT divisions. When the Fort Worth and Denver laid tracks across the XIT in 1888, general headquarters was moved from Alamositas to Channing, a townsite laid out by the ranch alongside the railroad. A phone line was installed at the new brick headquarters building, and by the early twentieth century numerous telephones were located on the ranch. Often the telephone line formed the top wire of a fence, with predictable service interruptions.

Each Christmas and New Year's XIT cowboys hosted all-night dances at the two-story Hotel Rivers in Channing, providing turkey, deer, antelope and, of course, beef. Alongside the fifty miles

Headquarters of the Rita Blanco Division, northwest of Channing.
—Courtesy Panhandle-Plains Museum, Canyon

Rock fences and terraces at Casas Amarillas.

The Yellow House Division in 1894; most of these buildings remain in use today.
— Courtesy Panhandle-Plains Museum, Canyon

Roundup crew setting up camp on the XIT range.
— Courtesy Panhandle-Plains Museum, Canyon

Built in Channing in 1890, the XIT office has been beautifully restored.

Rings for horse reins, set in concrete outside the XIT office.

Vault door in the main room of the XIT office in Channing.

Las Escarbadas Division bunkhouse, built in 1886 near the New Mexico border.

of railroad right-of-way across XIT lands, ranch employees planted millet, sorghum, and vegetables so that travelers could see the farming potential. Toward this end a several-hundred-acre farm was maintained at the Rita Blanco Division.

Rustling was a serious problem on the XIT, especially from "maverickers" who stole large calves as yet unweaned, then stamped their own brands upon them. The Escarbada Division was especially plagued by rustlers. Located across the west-central expanse of the XIT, the Escarbada contained creek beds, ravines and scrub timber, and a long north-south fence proved to be inadequate separation from an isolated and lawless stretch of New Mexico.

In 1895 Ira Aten, sheriff of Castro County and a former Texas Ranger who had fought in at least five shootouts, was placed in charge of the Escarbada. He employed a pair of fellow ex-Rangers, as well as other tough gunmen, to ride the New Mexico fence line, (which sometimes was pulled down by rustlers boldly driving a stolen XIT herd west). For ten years Aten battled cattle thieves, taking care to double his life insurance, stay beyond the flickering light of campfires, and paint his residence windows dark green. Aten and other XIT men were instrumental in ridding the western Panhandle of outlawry.

During the late 1880s the XIT began to drive steers to fattening ranges in the north, a practice followed by a number of big

Texas ranches. From 1890 through 1896 the XIT drove from 10,000 to 20,000 steers annually to Montana headquarters sixty-five miles north of Miles City on Cedar Creek, the fine range "between the rivers" (the Yellowstone and the Missouri). Normally, the XIT sent five herds on the 850-mile journey; each herd was composed of 2,500 steers and driven by eight cowboys, a trail boss, a horse wrangler, and a cook. The journey took three months, during which time the trail hands were paid $35 a month (the XIT in Texas paid just $25 monthly).

Although most of the trail hands were laid off in Montana, enough men were retained to drive the combined remuda back to Texas. Perhaps the XIT's most famous animal was "Dunnie," a buckskin mustang so dependable on night herd that he was used season after season on the long drives. One of the five chuckwagons was used to feed the remuda crew during the two-month return

THE WOLFERS

A recurring problem in the Panhandle, especially in broken country such as the Caprock Escarpment and the breaks of the Canadian River, was caused by large wolves. These lobos killed about seventy-five head of cattle annually.

During the first few months of each year, while cow work was slack, laid-off cowboys were offered bounties of five to ten dollars per dead lobo. The XIT furnished each pair of wolfers a wagon, camp supplies, and saddle horses.

A wolf that had just gorged itself was easy to run down, but otherwise wolves were fast and could be caught only after a chase of ten or fifteen miles. Although litters of pups, sometimes numbering ten or more, offered the best return, wolfers had to crawl several feet into a black den, shoot the bitch, then somehow fish out the pups.

A successful wolfer could earn as much in a few months as his cowboy pay the rest of the year. In 1896, for example, veteran XIT wolfer Allen Stagg killed eighty-four lobos along the Canadian River.

journey, while the four "empties" were hitched together in two pairs and driven back by one team per pair.

The XIT "double-wintered" steers bred in the Panhandle, expanding them in loin and frame on the Montana range before selling them as mature steers on the Chicago market. The XIT drove its final herd north in 1897. After that date settlers and their fences closed the Montana Trail, making it necessary to pay railroad rates to move steers to Montana.

In 1895 A.G. Boyce retired as general manager of the XIT, after eighteen years of running Texas' largest ranch. He was replaced by H.G. Boice, another experienced cattleman, but the Capitol Syndicate had already begun selling off its lands. Cattle prices were in steady decline, while farmers and developers clamored for land. Opening a land office in Farwell, the XIT became a large-scale colonizer and sold most of its land by 1912. By that time cattle, horses, and equipment had been disposed of, and the remaining 350,000 acres were leased — and eventually sold — to farmers and ranchers. The heyday of the XIT was a scant two decades, but it lasted long enough to earn a permanent place in western lore as the frontier's largest cattle ranch.

✓ THE MATADOR

"The Ranche seems to me to be quite capable of carrying at least 80,000 cattle. . . .It is without a doubt one of the best watered, sheltered and healthiest Ranches in Texas." This glowing description was part of a prospectus prepared in 1882 by Thomas Lawson for fellow Scotsmen hopeful of investing in the western "Beef Bonanza."

The attractive "Ranche" recommended by Lawson was the Matador, organized in 1879 to the east of the Caprock Escarpment in West Texas on open range of more than a million and a half acres. The Matador would indeed prove capable of grazing 80,000 head of cattle, and over a period of seven decades the Scottish stockholders provided stable, farsighted leadership unmatched by any other British ranching syndicate.

In 1878 Joe Browning drove a small herd of cattle to the heart

Matador cowboys around chuckwagon in the 1880s.

J.W. "Waddy" Peacock, an old-time Matador hand.
— Courtesy Panhandle-Plains Museum, Canyon

Matador cowboys in 1924. Given to the author by Herb Collins, at far right.

of what would become the Matador. Browning staked claim to a half-section of land at Ballard Springs, where a buffalo hunter had built a dugout at the eventual headquarters site of the Matador. Other herds soon were driven into the area, tended by men who lived in other dugouts and claimed the customary grazing rights of the open range.

Also in 1878 an experienced Texas traildriver, Henry Harrison "Paint" Campbell, put together a herd of his own which he sold at a handsome profit in Chicago. A Chicago banker, Col. Alfred Markham Britton, was impressed by Campbell's success and his knowledge of the cattle business. Britton persuaded Campbell to ramrod a Texas ranching operation to be called the Matador Cattle Company. Britton, his brother-in-law, Campbell, John W. Nichols, and Bud Lomax each put up $10,000 (Lomax, who was interested in Spanish culture, suggested the name "Matador").

Backed by Matador capital, Campbell returned to Texas, purchased a small herd of cattle, then headed west to locate a promising range. At Ballard Springs the dugout and range rights of Joe Browning were purchased, and Campbell established Matador headquarters in the dugout. The Matador Cattle Company was incorporated under Texas laws in 1879, and in December John Dawson sold the growing ranch 1,300 cattle branded with a $\mathcal{V}$ on the right side. Many other herds with many other brands were purchased during this formative period, but the $\mathcal{V}$ would become famous as the mark of the Matador.

Cattlemen who sold their herds to the new ranch usually surrendered their range rights as well. The Matador Cattle Company also acquired legal title to more than 100,000 acres, and soon Matador livestock were herded by line riders on an unfenced range of one and a half million acres. Henry Campbell had materials hauled in by wagon to erect a two-room plank ranch house at Ballard Springs. In 1881 Campbell drove a herd south to the Texas and Pacific railhead at Colorado City, receiving $75 per steer for the first Matador cattle sale. By this time Matador riders branded between 15,000 to 20,000 calves per year.

By this time, too, the Beef Bonanza was in full boom, and the Matador owners determined to try to take advantage of the lucrative prices being paid by British investors for western cattle ranches. Alfred Britton introduced the Matador to a group of business-

MERRY CHRISTMAS, MATADOR STYLE

Mrs. H.H. Campbell, the lovely and capable wife of the Matador manager, made the ranch a social center of lonely West Texas. She happily organized dances, church services, and other activities for the cowboys, and riders from adjoining ranches.

The highlight of the year was the Christmas dance, begun in 1882 at the two-room house at Ballard Springs. Mrs. Campbell assembled five other women from as far away as one hundred miles, and headquarters cook Ben Brock and Jim Browning provided fiddle music. With fifty cowboys waiting in line to dance with the six women, the festivities went on for two nights.

By the next year there was a large bunkhouse and a stone mess hall, as well as more women, and in future years the two-room residence was expanded into the "White House." Mrs. Campbell soon entertained upwards of one hundred people, but she demanded that there would be no liquor and no quarreling. During the summers Mrs. Campbell put up gallons of wild plum jelly to take the place of cranberries. As Christmas approached, some of the cowboys hunted deer, antelope, and turkey. A beef or two was slaughtered and barbequed, and two days of baking produced dozens of cakes and loaves of bread. There were tubs full of doughnuts and hundreds of fried dried apple pies, as well as gallons of black coffee.

On the afternoon before the first night, guests began arriving. Matador cowboys who had been laid off for the winter always were in attendance. The women brought their party clothes in a suitcase and changed in the White House.

The party began with a supper in the mess hall; for the remainder of the two nights and a day food would be available at a serve-yourself buffet set out in the bunkhouse. When the opening supper ended the tables were removed from the mess hall, the fiddlers tuned up, and the caller directed, "Get your partners!"

The dancing went on for the next thirty hours. Women were always outnumbered at least three or four to one by the men and had a constant line of expectant partners. The men swung their partners so enthusiastically that their exertions were minimized, and occasionally they would slip away for a nap in the White House. Periodically, the cowboys would snack in the bunkhouse or, in discreet defiance of Mrs. Campbell's temperance edict, nip from a jug.

men from Dundee, Scotland, and they employed Thomas Lawson to investigate the property and submit a prospectus. In 1882 a joint-stock company, the Matador Land and Cattle Company, Limited, was incorporated with offices in Dundee. Britton agreed to serve as American manager and Henry Campbell continued to function as ranch superintendent. Britton, Campbell, and their associates were paid $1.25 million for the Matador, including 40,000 head of cattle, 100,000 acres of land, "and range and other rights and privileges in or over 1,500,000 acres. . . ." (Britton and Lomax immediately set about organizing another large ranch — the Spur, just southwest of the Matador — to sell to another British syndicate.)

The investors who comprised the Matador company were successful businessmen who also had a solid grasp of animal husbandry, based upon ancient Scottish practices of cattle raising. The board of directors demanded detailed reports from the ranch and made decisions at monthly meetings in Dundee. The secretary of the board, Alexander Mackay, served the company skillfully for fifty-four years, until his death at the age of eighty in 1936. Mackay made annual trips to the ranch during the fall roundup season, usually accompanied by a board member. The board thus became closely acquainted with their ranch, while Matador employees gained a personal sense of foreign ranching syndicates. Careful calculations of annual operating expenses, as well as debt retirement obligations, determined board instructions regarding cattle sales for each year.

The board quickly decided to obtain legal title to as much land as possible, and purchases of large acreages became regular occurrences. Fencing and corral construction was another regular expenditure. The first fence was fifty-one miles long across the northern boundary. In 1884 a fence was put up to separate the Matador from the Spur Ranch to the southwest, and the first fenced subdivision was a forty-section horse pasture near the Ballard Springs headquarters. A dam construction at headquarters created a spring-fed lake that was stocked with fish, and tanks and windmills were erected throughout the vast range.

The Matador range was broken and brushy. Hereford bulls were introduced by the hundreds to the Matador herds. Within a few years 500 of the best male calves born annually to Matador cows were kept as future sires. The ranch maintained four bulls per one hundred cows, and during good years the calf crop was as high

Built about 1916 as the Mess House, this stone structure at Matador headquarters today is an employee residence.

This spring-fed stone water tank towers above the ranch site, providing Matador headquarters with running water by gravity.

Built about 1880 as a guest house for absentee Scottish owners, in the 1930s this frame structure became the business office for the Matador Land and Cattle Company. Today it is part of the Ranch Heritage Museum in Lubbock.

The Quanah, Acme & Pacific depot, built in 1913, is the oldest structure in Roaring Springs.

as eighty percent. The sturdy Matador cattle which evolved became adept at hiding in the brush and earned a reputation for wildness.

The Matador horse herd was developed through a Palomino stallion, Peter McQue, and his son, Shiek. These studs were bred with thoroughbred mares and mares showing Morgan blood, in time resulting in a strong, heavy-bodied cowhorse well-suited to the rugged Matador country.

The position of ranch manager became vacant in 1890, and the board selected a man destined to become legendary among cattlemen. Born in Scotland in 1850, Murdo Mackenzie acquired experience in law and banking. At the age of thirty-five he came to the United States as manager of the Prairie Cattle Company, headquartered in Edinburgh and centered in southeastern Colorado and northeastern New Mexico. Mackenzie and his family made their home in Trinidad, Colorado, and when he was employed by the Matador Company five years later, he insisted on moving the American office from Fort Worth to Trinidad. Aside from wanting to maintain his home and friends, Mackenzie felt he would be in a better position to travel to the Chicago market and to supervise the Matador's northern ranges.

Like other large Texas ranches, the Matador had begun driving two-year-old steers to richer northern grasslands for double-wintering before sale. Through the years the Matador would lease hundreds of thousands of acres of range in Montana, South Dakota, and Canada. During the 1890s the Matador also expanded into the Texas Panhandle, purchasing the 214,000-acre Alamositas Division from the XIT and leasing an even larger pasture near White Deer.

Soon after assuming leadership of the Matador, Mackenzie issued orders that employees would not gamble on the ranch or trail drives and "to strictly prohibit any of our men from frequenting the Saloon." Many cowboys resented these restrictions, and

Murdo Mackenzie (1850-1939) managed the Matador for thirty-six years.
— Panhandle-Plains Museum, Canyon

Mackenzie's life was threatened. Not wanting "to show the white feather," Mackenzie fearlessly went about his duties. "If, however, I had known what I had to go through I would never have undertaken it," he later admitted.

Mackenzie spent more time traveling on business than enjoying his Trinidad home. On his tours of Matador pastures he drove a buggy and team around the range. When the American Stock Growers Association was organized in 1905, the widely respected Mackenzie was the overwhelming choice for first president.

By that time the original Matador spread consisted of 800,000 acres, on which 70,000 cattle regularly grazed. Three dozen employees worked through the winter months, but the work force expanded to at least seventy-five during the roundup seasons. On the other Matador ranges, of course, there were more cattle and cowboys. Range detectives also were employed to discourage rustlers and trespassers. Legal expenses were considerable, because the company had the resources and inclination to protect its interests in court.

In 1909 Dode Mackenzie, son of Murdo and manager of the Dakota Division, was shot to death by a disgruntled former employee. The tragedy may have inclined Murdo to seek new surroundings. In 1911 he accepted a lucrative offer from the Brazil Land, Cattle, and Packing Company to move to Sao Paulo and manage their 2.5-million-acre operation.

Mackenzie was succeeded by another native of Scotland, John McBain, who had served as bookkeeper at Ballard Springs since 1898. The community of Matador, seat of Motley County, had grown up two miles north of Ballard Springs, but when a railroad finally crossed Matador range in 1913, tracks were laid well to the south of town. The Matador paid the railroad company the sum of $85,000 for building on the ranch, while laying out the townsite of Roaring Springs along the tracks eight miles south of Matador. The citizens of Matador soon backed construction of an eight-mile short line to their town.

During World War I, 1914–1918, the Matador enjoyed a peak sales period. The inevitable postwar decline brought cost-cutting to the Matador. Among other economies, cowboy wages were cut from $50 to $40 a month for old hands while new men were hired for less. During the 1920s Matador management encouraged oil explorations on their property; however, unlike other vast Texas ranches, significant discoveries never benefited the Matador.

Murdo Mackenzie returned from Brazil in 1917, took up residence in Chicago, and accepted a position on the Matador board. When John McBain died unexpectedly in 1922 in Denver, where he had moved the Matador offices, Mackenzie volunteered to serve as a replacement. The seventy-two-year-old Mackenzie accepted appointment as manager at an annual salary of $10,000, then moved from Chicago to Denver. Mackenzie soon accepted the position on a permanent basis, appointing his son, John, as assistant manager.

By the late 1920s, the growing public preference for baby beef, combined with increased expenses of the northern divisions, produced the policy of pulling back to the Matador and Alamositas divisions in Texas and selling two-year-olds. In 1936 Alexander Mackay died suddenly in Scotland, and the next year the eighty-seven-year-old Mackenzie retired to his Denver home. John Mackenzie was promoted to his father's position, while Murdo died in Denver in 1939. In 1940 John V. Stevens, an experienced cowboy who had graduated from Texas A&M, was appointed manager of the Alamositas Division, then placed in charge of the Matador Division in 1946.

With sound business policies and strong financial backing the Matador continued as a profitable operation, the sole foreign ranching syndicate which had survived from frontier days. But in 1951 the board accepted a lucrative offer from a New York group: seventy-cent shares would be priced at $23.70. Involved in the sale was a cattle herd of more than 45,000 animals, 1,400 horses, the 400,000-acre Matador Division (considerable acreage had been sold to farmers through the years), and the Alamositas Division, which had been expanded to 395,000 acres. During the 1950s the historic old ranch was disposed of in parcels ranging from 16,000 to 96,000 acres. A substantial ranching operation still is centered around the handsome Matador headquarters at Ballard Springs, and John V. Stevens, now (1997) in his eighties, continues to run a small ranch that once was Matador property.

Harry P. Drought of San Antonio, who had become a Matador board member in 1938 and who possessed a strong sense of tradition, remarked upon the liquidation of the great ranch: "Regardless of how advantageous to the shareholders the sale may have been,

One of the earliest structures at Matador head-
quarters. By the twentieth
century two employees
lived here who were
assigned to keep head-
quarters tidy.

The main house at the
Matador was built atop
a hill just south of the
town that was named
for the ranch.

The original office at the
Matador was known as
the "Scotch dive."

there are many heartaches caused by this conclusion. The Company, however, will live forever in the history of the Southwest. We were connected with a cattle empire. . . ."

⅄ SPUR RANCH

"This virgin range was all open country, and I have never seen its equal," recalled Frank Collinson, the first cattleman to lead a herd into the country soon to become the Spur Ranch. "Water was good; grass and wood were plentiful. There were deer in every thicket and antelope on every flat. Plums, currants, and grapes, the finest that ever grew, were plentiful."

This alluring range lay just east of the base of the Caprock Escarpment, a long north-south fault line which thrusts upward to the high plains of West Texas and the Panhandle. Numerous springs flowed from the base of the Escarpment, feeding streams in the rolling terrain to the east of the Caprock. In addition to a dependable source of water, this broken country provided protection from winter blizzards. The region teemed with buffalo, until professional hunters annihilated the herds during the 1870s; so many bones littered the ground that pioneer ranchers hauled them by wagon to the nearest railroad to sell for $6 per ton.

With the disappearance of buffalo herds and the warlike Comanches who subsisted upon them, cattlemen quickly preempted the free range. Frank Collinson drove in the initial herd, purchased in 1878 from John Chisum in eastern New Mexico. Soon there were thirty cattlemen running small herds on free range.

The first herd in the region to use a Spur brand was trailed in from New Mexico by J. M. Hall. In 1882 Hall sold his cattle and the Spur brand to Tom Stephens and Coleman Harris, who sold the brand and herd the next year to the Espuela Cattle Company. *"Espuela"* is the Spanish word for "spur," one of the most common brands in Texas. This corporation was composed of A.M. Britton as president, S.W. Lomax as secretary, Tom Stephens, S.T. Pepper, and Bud Lomax. Two years earlier Britton and Lomax had formed the Matador Ranch to the north, then sold the spread for $1.25 million in 1883 to a Scottish syndicate. Hoping to repeat this success in the

Spur drovers returning home by rail after a long drive.

boom conditions of the Beef Bonanza, Britton and Lomax immediately went to work organizing a new ranch promotion.

When the Lone Star Republic joined the United States in 1845,
the new state of Texas retained possession of public lands (public
lands in all other states were owned by the federal government).
Later, Texas offered railroad companies sixteen sections of land for
every mile of track laid within the state. The Houston and Great
Northern Railroad, which constructed 215 miles of track in southeastern Texas, was entitled to 3,440 sections of land, available in the
western part of the state. The Espuela Cattle Company paid
$515,440 for 378 sections in Kent, Dickens, Crosby, and Garza
counties, all of which are square in shape and have a common
boundary point. The state retained alternate sections in all railroad
blocks for school lands, and the Espuela Cattle Company leased
these sections at a nominal rate.

The first modest headquarters of the Spur Ranch was on Red
Mud Creek in southern Dickens County. Eventually, the Spur
Ranch fenced in a total of 569,120 acres. In 1884 the Espuela Cattle

Company changed its name under Texas corporation laws to the Espuela Land and Cattle Company, although the outfit would commonly be known as the ranch or the Spurs. The Spur Ranch was stocked in part by buying the herds already grazing on Espuela lands. Some of the small ranchers sold out for cash settlements, but others took part cash and part stock in the Espuela Land and Cattle Company.

More numerous stockholders were found in England and Scotland, where A.M. Britton traveled to find investors. Wealthy British capitalists organized the Espuela Land and Cattle Company of Fort Worth on April 9, 1885. In addition to the large British investors, there were hundreds of small stockholders — pub owners, clerks, teachers, ministers — who held as few as ten shares.

In the summer of 1885 representatives of the London company arrived in Texas to take title to the Spur Ranch and to count the cattle. They were fortunate in making the acquaintance of Charles Goodnight, already a legendary cattleman, who loaned them his foreman, John Farrington, and two experienced cowboys to assist with the tally. The Fort Worth company was represented by Bud Campbell, who then became boss of the Spurs.

The Spur Ranch already had been enclosed by a barbed wire fence, creating a half-million-acre pasture. Farrington suggested putting up a north-south cross fence, and all Spur cattle were driven into the East Pasture. Then they were rounded up and moved into the West Pasture, providing an accurate tally as they passed through

Spur cowboys.

UMBRELLAS, BOWLERS AND . . . COWLETS?

In 1889 John McNab, a wealthy Scotsman who was a member of the Espuela board of directors, made the first of several visits to the Spur Ranch. The day after McNab arrived a roundup was scheduled in the vast East Pasture, and he announced a desire to attend. A gentle horse was saddled and brought up for McNab.

"No, thank you," said McNab. "I will walk."

McNab could not be dissuaded, and he headed for the East Pasture dressed in proper London attire: business suit and cravat with starched collar and cuffs; stiff bowler hat; polished shoes; and an umbrella, which was not needed for rain protection but provided welcome shade from the withering Texas sun. Half-wild Spur cattle, however, became skittish at the unfamiliar sight of a pedestrian with an open umbrella, and when McNab was around the cowboys went on alert to head off stampedes. The cowboys were asked by ranch manager Fred Horsbrugh not to shoot holes in McNab's umbrella or bowler.

On future visits McNab occasionally agreed to ride in a buggy, if Fred Horsbrugh drove, but he preferred to set out on foot. Since he usually became lost when he strayed very far from headquarters, a cowboy always rode out of sight to "loose herd" McNab. When McNab became bewildered, the cowboy would appear as if by coincidence and offer to guide him.

McNab once brought his wife, along with Mr. and Mrs. Alexander Mackay of the Matador Land and Cattle Company of Dundee, Scotland. The women watched a roundup at which bull calves were being branded, earmarked, and castrated.

"It is terrible to treat those little cowlets like that!" exclaimed Mrs. Mackay.

"Ma'am, them's not cowlets," remarked a cowboy with a straight face. "Them's bullets."

the gates. The herd was considerably short of the 40,000 head promised by the Fort Worth company, which had to purchase enough cattle to make up the difference.

The ranch also bought about 800 horses, and through the years tried to keep at least ten horses per cowboy. Unlike various other ranches, the Spurs raised no horses; all mounts were purchased. The Spurs generally paid about $5 per head extra ($35 instead of $30 was a common price) for "broken" mounts, so that the ranch would not have to go to the expense and trouble of training unbroken horses. The original Spur dugout headquarters became a complex of plank buildings, but dugouts were maintained around the ranch for line camps and as homes for married hands.

S.W. Lomax was appointed manager of the Spur Ranch. Dignified, well educated and widely traveled, Lomax was greatly respected by his men. Mrs. Lomax had no intention of living on an isolated ranch, so she stayed with their five children at a handsome home in Fort Worth. (The absent Mrs. Lomax arranged no social life for Spur cowboys, a dull tradition that continued throughout the history of the ranch. Fortunately, the Matador Ranch to the northeast organized numerous social events, and Spur cowboys were always invited.)

Although the manager reported to the board of directors, he was given power of attorney and almost complete discretion in running the ranch. In London the board secretary was a full-time employee devoted to affairs of the ranch, and Englishman S.G. Flook, nephew of a board member, was sent to the Spurs as bookkeeper. In 1886 Scotsman Fred Horsbrugh secured a position as assistant manager of the ranch, partially to keep the board informed of any mismanagement. Even though Lomax kept meticulous records and was regarded in Texas as an excellent ranch manager, the Spurs was plagued during the 1880s by drought and low cattle prices. In 1889 the board tried to cut Lomax's $7,500 salary in half, but he resigned and accepted an offer from a bank at Vernon. Fred Horsbrugh became the new ranch manager — at $2,000 per year.

Horsbrugh remained in charge of the Spurs for fifteen years, but it proved difficult to produce profits for the British stockholders. The herd was increased to more than 54,000, and upgrading was begun by purchasing 200 Hereford and Shorthorn bulls annually for six years. During the roundups the best bull calves were selected for eventual breeding with the finest Spur cows, and by 1895 most of the herd had white faces.

But range conditions often were poor because of frequent drought or prairie fires. Cattle weakened by inadequate grazing died by the thousands when winter blizzards struck. During dry spells, cattle that had to travel miles to water were weakened because they spent so much of the day walking instead of grazing. Such conditions often caused a serious drop in the year's calf crop. Heel flies sent cattle scurrying to water, where they stood protecting their heels instead of grazing; during heel fly season bog riders constantly had to pull out cattle mired in mud or quicksand, and if they were too late the beasts died. Hailstorms sometimes dropped stones big enough to kill large numbers of calves, and calves also were the frequent prey of wolves.

Aside from regular losses from these causes and from cattle thieves, the Spur Ranch often suffered from depressed markets. Income from cattle sales sometimes fluctuated wildly, most notably in 1892 ($109,800), 1893 ($16,826 — the Espuela's all-time low), and 1894 ($188,328 — the all-time high). Although the Spurs long maintained a large pasture to the north near White Deer, the board never approved a finishing range in Wyoming or Montana. Other big Spur herds were finished before sale on the frequently inadequate West Texas ranges owned by Espuela. Numerous tanks were built, and fifty-seven windmills eventually improved the Spurs water supply, but a far greater investment was needed.

The largest annual expense on the Spur Ranch was wages (the second largest was taxes, including a high of $17,282.05 in 1900, and the third largest was supplies, averaging a little more than $4,000 per year). The greatest expenditure for "hands" (Spur records never referred to the men as cowboys) occurred in 1887, when $25,731 was paid to cowboys and other personnel. The ranch averaged seventy-two employees per month that year, although there were far fewer hands during the winter months, and many more during the spring and fall roundups. The average in 1888 was sixty-six, but during the first few years crews were busy constructing fences and tanks, and by 1889 the average dropped below forty-nine. The lowest employee average was twenty-five, in 1896, and drought and poor cattle prices drove the wages per man down; in 1900 the employee average was thirty-six, but wages paid totaled merely

PINK HIGGINS
VS. BILL STANDIFER

Plagued with cattle thieves on the vast Spur range, in 1899 manager Fred Horsbrugh employed two special range riders — a hardcase named Tynam, and the former sheriff of Crosby County, Bill Standifer. Tynam soon moved on, but was replaced by Pink Higgins, a veteran cattleman and gunman who had won special notoriety during a bloody range war against the Horrell brothers in Lampasas County.

The range riders proved "very successful in putting down stealing," according to Horsbrugh, but Standifer and Higgins clashed over "an old grievance." When Horsbrugh could not stop the quarrel, he dismissed the two gunmen in August 1902, although he agreed to let Higgins remain in his house until Pink could arrange to move where his children could attend school.

But Standifer returned, and a "premeditated (and naturally arranged for) dual [sic]" took place within sight of Pink's house on October 4, 1903. The two antagonists rode toward each other, dismounted when sixty yards apart, then began exchanging rifle fire. Higgins' horse was struck, but Pink scored with a bullet that tore through Standifer's elbow and into his heart. Standifer shifted his rifle to the crook of his good arm, started to stagger away, then fell dead.

It was a fair fight, and Higgins was acquitted following a quick trial at Clairemont, the seat of Kent County. Having lived up to his fearsome reputation, he was appointed a deputy sheriff of Kent County — and rehired as a range rider by the Spur Ranch.

$8,967. Within another year improving conditions sent the employee average up to nearly forty-nine, and the average remained well over forty during the last several years of the Espuela Company's existence.

The monthly wage scale ranged from $30, for an ordinary hand, to the highest paid employee (range foreman) at $125. Spur hands earned their pay, working seven days a week, with three days at Christmas the only holiday of the year.

Fred Horsbrugh struggled to save money on every activity and

The Spurs employed two range riders to dissuade rustlers. But Billy Standifer and Pink Higgins developed hard feelings, and Standifer was killed in a shootout on Spur range.

transaction. Profits proved impossible to produce, however, and the board asked for Horsbrugh's resignation in 1904. The able and popular Horsbrugh was replaced by Henry Johnstone, a Scottish relative of founders of the company, but at this point the board began trying to sell their unprofitable property. Horsbrugh acted as an agent and interested the S.M. Swenson Company in the Spurs. Transactions were completed in 1907, with the Swenson syndicate acquiring title to 437,670 acres of land and 30,000 cattle and horses. The new owners laid out the townsite of Spur and other communities, promoted a railroad into Spur, and began selling parcels of land to farmers and small ranchers. A large ranch was maintained in part of the old West Pasture, but the Swenson SMS brand was used on the cattle.

The Espuela Land and Cattle Company had been organized because of the Beef Bonanza, but by the time the Spur Ranch began large-scale operations the boom conditions had entered reversal. Although the company made substantial financial commitments, investment limits were drawn on water improvements and northern finishing ranges which might have made the Spurs a profitable enterprise. The Espuela Land and Cattle Company, entering a tough business in a harsh area at a bad time, proved consistently unprofitable for more than two decades.

Despite its ups and downs, the ranch's hands remained loyal. A prime example was Jack Raines, who signed on when the Spurs was taken over by the British company and worked steadily until Espuela sold out in 1907 to the Swenson interests — then went on the Swenson payroll, and rode the same pasture in the 1930s that he had ridden in the 1880s.

Office buildings erected in Stamford for the Swenson Land and Cattle Company.

The Spur granary and stable, built in 1906 on the side of the hill. Now located at the Ranch Heritage Museum in Lubbock.

THE WOLF BOUNTY ACCOUNT

Wolves were such a problem on the Spurs that a special Wolf Bounty Account was carried on the ranch books. Any cowboy, professional hunter, or settler was offered $5 for lobo scalps and fifty cents for coyote scalps (coyotes were of more danger to sheep than cattle). County governments also paid bounties, and fifty to sixty lobos and as many as 300 coyotes were killed annually in Spur pastures.

In June 1894 Spur cowboys Will Monroe and C.R. Humphrey tracked a female lobo to a den. The hole was big enough to crawl into, and Monroe, the smaller of the two, wriggled inside wearing heavy leather gloves. Suddenly, a big she-wolf bolted from a side compartment and ran out between Humphrey's legs, carrying the startled cowboy on her back for several strides.

Apparently, several bitch wolves used the den; Monroe handed out fifty-three cubs from several compartments. Larger pups were scalped, while the smaller ones were gutted and brought in with scalps intact for evidence. Monroe and Humphrey were paid $265 by the ranch and the same amount by Dickens County — more than half a year's wages for each man.

Ψ THE PITCHFORK LAND AND CATTLE COMPANY

Organized in 1883, the Pitchfork is one of the few ranches from the golden era of the Beef Bonanza still in operation, and the only one that is larger today than in its frontier heyday. The Pitchfork still is owned by the St. Louis family that put up most of the founding capital, even though the ranch is one of the rare big Texas spreads not to benefit from substantial oil production.

Late in 1881 Dan Gardner, who had ridden on trail drives, served as a ranch foreman and helped to found the Texas and Southern Cattle Raisers Association, learned of an opportunity to buy an outfit owned by Jerry Savage. There were 2,600 longhorns bearing a Pitchfork brand, seventy horses, wagons and camp equipment, a little deeded acreage, and range rights along the South Wichita River in eastern Dickens and western King counties. Gardner lined up a former employer to provide most of the $50,000 purchase price, but the man was forced to withdraw suddenly. Gardner arranged a meeting in Fort Worth with Eugene Williams, a distant cousin and boyhood companion who had become vice-president of a St. Louis shoe company.

Williams agreed to furnish four-fifths of the money, then the two promptly began expanding their ranch. In 1883 Williams sold half of his four-fifth interest to another St. Louis businessman, A.P. Bush, Jr., for $50,000, and this influx of capital helped increase the Pitchfork herd to 6,000 by the end of the year. By this time, however, the big Matador and Spur ranches had been organized nearby, and a merchant named Samuel Lazarus had acquired deeds or leases to 100,000 acres, much of which lay within the pre-empted range of the Pitchfork. Lazarus placed 3,750 head of cattle on his land, but the Pitchfork owners quickly came to a partnership agreement with him. Williams enlisted

The distinctive Pitchfork symbol on a pasture gate.

two of his St. Louis associates as investors, and the six businessmen and ranchers incorporated the Pitchfork Land and Cattle Company on December 29, 1883. Lazarus soon sold his interest back to the company, which now had a combined herd of nearly 10,000 cattle, and Dan Gardner was named ranch manager.

"Everything getting along splendidly," reported Gardner to his fellow directors in May 1884, "except for a case of measles and a man's getting his finger pulled off a few days ago." A case of measles was a bit unusual, but a finger jerked off while dallying a loop around a saddle horn was an occupational hazard for cowboys.

For years the 'Forks employed nine cowboys during the winter and about twenty during roundups and branding. By 1886 the herd had grown to around 17,000, and the next year Gardner initiated a farming operation to produce supplemental feed, at first for horses but soon for the cattle. Title was acquired to more acreage, while a fencing program enclosed the 97,000-acre Pitchfork range by 1896. Dugouts scattered around the 'Forks provided the only living quarters for about a decade and a half. In 1886 a three-room dugout was cut into the river embankment at the eventual headquarters site, and eleven years later a three-room frame house was constructed. The house was moved to higher ground in 1903 because of flooding, joining other frame structures that at last began to give a permanent look to headquarters.

Eugene Williams traded his stock in the St. Louis shoe company for Pitchfork stock, making him the majority shareholder in the ranch. The shoe company later went bankrupt, but William fell into ill health and died at the age of forty-nine in 1900. His two sons were too young to take an active role, leaving Dan Gardner in virtual control of the 'Forks. The spread soon became known in West Texas as "Dan Gardner's Pitchfork Ranch."

Gardner had married Sula Ellison in 1887 and they made their home in Fort Worth, where the Pitchfork maintained an office. But in 1889 Sula died giving birth to a son, Sula Ryan Gardner. The boy was reared in Fort Worth by his maternal grandmother, while the bereaved father moved back to the 'Forks. He continued to travel frequently to the Fort Worth office, and he made numerous other trips common to managers of large ranches.

Gardner bought the Pitchfork's first automobile, a Hupmobile, about 1910, using this car and later autos to tour pastures

COYOTE COUNTRY

In the 1930s, after half a century of existence, the Pitchfork range was more troubled by coyotes than during its pioneer period. The coyotes constantly preyed on calves and, with government assistance, the ranch employed trapper Jack Beachly in 1936. In January he captured sixty-six coyotes, with three bobcats for good measure. Within three more months Beachly had trapped 400 coyotes, and for the next three years he averaged snaring one coyote per day. The predator population was considerably reduced, but to this day coyotes still slip up behind the cookhouse searching for scraps.

as well as to facilitate his constant travels. A telephone was installed at headquarters about 1903, but the line was atop a fence and for many years service was unreliable. More reliable was a radio, purchased in 1923 for immediate cattle market information as well as for entertainment at the isolated headquarters. A thirty-two-volt engine brought to headquarters in 1917 proved inadequate for lighting the cookhouse, two-story bunkhouse, and several residences, but a larger generator provided electricity for thirty years, until REA lines at last were constructed to the ranch in 1948.

Dan Gardner was elected company president in 1914; Georgie Williams, widow of the late Eugene Williams, was chosen vice-president; and the Williams sons, Eugene and Gates, became increasingly active in management of the ranch as well as other family financial affairs. When Gardner died at the age of seventy-seven in 1928, the Williams brothers assumed a much larger role in managing the Pitchfork. Ranch headquarters was extensively remodeled, the "Big House" was handsomely expanded, and a landing field was cleared as a transportation convenience for Gates Williams.

Following an exhaustive search for a ranch manager, the Pitchfork signed Virgil Parr, a Texas A&M graduate and a nationally known specialist in agricultural economics. The new manager cross-fenced the 'Forks into pastures of 4,000–6,000 acres, greatly enlarged the farm, and modernized housing at the outlying camps. Parr noticeably improved the cattle and horse herds, as well as the

The chuck box from an old Pitchfork chuckwagon. The ranch still maintains a horse-drawn chuckwagon for roundups in the rugged south range.

Side view of the old Pitchfork office, with a buggy parked beside it. The porch of the new office is at left.

The old office at the Pitchfork has been converted into a museum.

The wall phone and wood-burning stove were fixtures of the old Pitchfork office.

windmills and tanks on the 'Forks. But Parr's improvement program was expensive; expenditures in 1927, for example, amounted to only $15,000, but in 1938 the total was $104,000. Relations between Parr and the Williams brothers became strained, and the manager turned in his resignation at the St. Louis board meeting in January 1940.

Parr was replaced by Rudolph Swenson, youngest son of the manager of the nearby Swenson Land and Cattle Company. But two years later an auto-train accident claimed the life of Swenson, and inflicted severe injuries on the daughter of Gates Williams. Douglas "D." Burns, a Texas A&M graduate whose brother managed the 6666 Ranch, was appointed manager in 1942. The Pitchfork enjoyed boom wartime markets over the next few years, but D. Burns had to deal with a severe manpower shortage during World War II and the Korean War.

Burns upgraded the Pitchfork Herefords, and it was decided to seek additional range for herd expansions. In 1947 the Pitchfork bought 43,000 acres from the Matador, along with another 15,000 acres in 1952 when the great ranch was broken up. Also in 1952 the Pitchfork acquired the historic Flag Ranch, seven miles south of Laramie, Wyoming. The Wyoming range proved invaluable during the long drought that wracked West Texas in the 1950s. When D. Burns retired at the age of seventy in 1965, the Pitchfork had 161,078 acres under fence in Texas, plus 35,000 acres in Wyoming — 75,000 acres more than when he took over in 1942.

Gates Williams died in 1964 and Eugene followed him in death two years later. But Eugene, Jr., had been groomed to succeed his father and uncle, just as assistant manager Jim Humphreys long had been trained to replace Burns. Humphreys began experimenting with crossbred cattle, while investing in the latest technology, including a helicopter for use in roundups.

The Pitchfork still utilizes scattered camps, although camp employees live at modern residences and are provided with two beeves annually. Cowboys on the 'Forks continue to use the "rope and drag" method during branding, and they gather in the chuckhouse for hearty fare prepared over the ancient wood-burning stove. There has been remarkable stability of management backed by an ownership family with deep financial resources and affection for their ranch.

Dinner bell outside the Pitchfork cookhouse. The bell rings at 5:45 A.M. and 11:45 A.M.

Well on the porch of the Pitchfork cookhouse.

The Pitchfork cookhouse can accommodate thirty-two hungry cowboys.

The ancient range in the Pitchfork cookhouse.

While most big frontier Texas ranches disappeared or shriveled to a fraction of their original size, the Pitchfork has survived and grown, well into its second century, into a larger enterprise with sales and expenses measured in the millions of dollars annually.

The main residence at the Pitchfork is called "The White House."

WAGGONER RANCH

"I see there is an oil well for every cow," cracked humorist Will Rogers during one of his visits to the oil-rich Waggoner Ranch in the 1920s.

Even before oil was discovered, Daniel and W. T. "Tom" Waggoner, father and son, had built a pioneer ranch into a million-acre operation with 60,000 head of cattle carrying their famous Three D brand. Born in Tennessee in 1828, Daniel Waggoner moved with his family to Texas in the mid-1840s. By the time Dan Waggoner was twenty-one he had learned the rudiments of cattle raising and trading from his father and had accumulated 242 longhorns, six horses, and a male slave. In 1851 he bought 15,000 acres on the Trinity River in Wise County eighteen miles west of recently founded Decatur. Comanche and Kiowa raiders temporarily drove him to the eastern part of the county, but he steadily increased his herd.

Dan married Nancy Moore in 1848, and in 1852 she gave birth to their only son, William Thomas. After the death of Nancy, Dan remarried in 1859, and he began to teach his son the ranching business. By the time Tom was seventeen, in 1869, his father had made him a partner.

After wintering a longhorn herd in Clay County, in 1870 Dan and Tom led a drive to Kansas, then rode back to Texas with $55,000

Waggoner commissary, built in Wichita County in 1870, but now part of the Ranch Heritage Museum in Lubbock.

in their saddlebags. The Waggoners used this capital to shift their growing operation westward, establishing a new headquarters at the junction of China Creek and the Red River in northwestern Wichita County. By the early 1880s they had extended their range thirty miles southward to the Pease River. But the Three D herd was expanding so rapidly that the Waggoners leased 650,000 acres of reservation lands north of the Red River.

The Waggoners began breeding Shorthorns by 1885, but Herefords were introduced in the 1890s and eventually predominated. Fine horses long had been bred by the Waggoners. Tom's favorite mount, Cow Puncher, was sired by a Palomino stallion quarter horse named Yellow Jacket. The most famous Waggoner quarter horse was Poco Bueno, a cutting horse champion and superb stud. Tom eventually poured a fortune into breeding thoroughbreds and constructing a racetrack at Arlington between Dallas and Fort Worth.

By the turn of the century, the Waggoner Ranch ran thirty miles east and west and twenty-five miles north and south, centered in Wilbarger and Wichita counties, with spillover in Archer, Baylor, Foard, and Knox counties. In 1903 the China Creek headquarters

THE THREE D BRAND

Dan Waggoner originally used a D brand, but around 1866 he began branding with three Ds in reverse. The Three D brand was easy to recognize and difficult to alter, but one daring rustler stole 144 steers and used a bar iron to box the Three Ds. The rustled steers were quickly located, and since no Box 3-D was registered Dan Waggoner recorded the brand in his name. Backed by several armed cowboys, Waggoner reclaimed his steers from the unprotesting brand burner.

was sold off as farm land at $15 per acre, funding the purchase of 100,000 acres of Wilbarger County land at $5 per acre. Headquarters was moved to Zacaweista, near Electra.

Ranch founder Dan Waggoner died in 1904 at the age of seventy-six. Tom moved to Fort Worth to supervise the Waggoner business interests, but continued to spend considerable time at Zacaweista and at his Decatur mansion. In 1909 Tom gave half of the 540,000-acre Waggoner Ranch to his three children to provide them experience in running a ranch. Paul, Guy, and Electra were presented with 90,000 acres each. However the discovery of oil transformed the nature and value of the property, and in 1923 the family holdings were merged into the Waggoner Estate.

Oil had been discovered accidentally in 1903 by a crew attempting to drill a deep-water well. "Oil, oil, what do I want with damn oil?" fumed Tom at this unexpected development. "I'm looking for water. That's what my cattle need."

But Waggoner soon adopted a more enlightened attitude as one gusher after another began pumping wealth into Waggoner coffers. Oil wells bearing the Three D brand inspired Will Rogers' remark about a well per cow. The Waggoner Refinery was built, and a chain of gasoline stations displayed the Three D brand. Enjoying one of the largest fortunes in the Southwest, Waggoner restored the family mansion in Decatur in 1931, the same year he built Arlington Downs racetrack. He and his wife, Ella, donated three buildings on the campus of Texas Woman's College in Denton, and in 1933 Tom

Dan Waggoner built "El Castile" for his growing family. Dominating a hill east of Decatur, the sixteen-room limestone mansion boasted five marble bathrooms, sixteen-foot-tall doors, and handsome isinglass fixtures from Denver.

was recognized as "first citizen of Fort Worth." He died the next year at eighty-two.

After Tom's death the Waggoner Estate, including ranch property, feed lots, farms, banks, oil operations and horse breeding stables, was administered by a trustee and a board of directors. With headquarters still housed in the handsome stuctures at Zacaweista, the Waggoner Ranch remains in the hands of descendants of the pioneer who began the family ranching operation before the Civil War.

THE RICH DADDY

In Decatur a church committee approached Tom Waggoner with a request for a donation to the building fund. He pledged $100, but a disappointed committee spokesman pointed out that one of Waggoner's sons had subscribed $1,000.

"Yes, I know," shrugged Waggoner, keenly aware of the free-spending habits of his offspring, "but the boy has a rich daddy."

THE RANCHER'S DAUGHTER

Electra Waggoner, the only daughter of wealthy oil and cattle baron Tom Waggoner, set all-time standards of extravagance. She owned 350 pairs of custom-made shoes, staged spectacular parties, and indulged in astounding shopping expeditions.

Born in 1882, Electra and her two brothers were raised in the palatial splendor of El Castile, the family mansion in Decatur. When a new townsite was organized near the Waggoner Ranch headquarters at Zacaweista, almost inevitably it was named Electra in honor of the pampered daughter. While touring Europe, Electra met A.B. Wharton of Philadelphia. When they married in 1902, her father built a mansion in Fort Worth, Thistle Hill, as a wedding gift. There also was a summer home in New Jersey, as well as shopping trips to Europe and the Orient.

But in 1921, after nineteen years of marriage, the Whartons divorced. By this time, oil money was pouring in from the ranch, and Electra found solace in a sustained spending spree. She bought a Georgian mansion in Dallas that covered three blocks. Christening the estate Shadowlawn, Electra furnished her magnificent home with splendid art objects collected during her travels and other finery totaling half a million dollars.

Electra's parties were legendary, with guest lists that included European royalty, movie stars, and miscellaneous denizens of high society. Weekend extravaganzas commenced at Shadowlawn, then the revelers piled into a caravan of Electra's cars and continued the entertainment at Zacaweista. Electra even arranged cross-country parties aboard a private train.

The large-scale merrymaking ended in 1922, when Electra remarried and sold Shadowlawn. She divorced two and a half years later, only to marry again within several months. The third marriage ended in annulment after a few months. By this time an overweight Electra was in failing health, and she died in a Park Avenue hotel suite in New York City on Thanksgiving Day. The spendthrift rancher's daughter was forty-three.

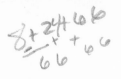

6666 Four Sixes Ranch

One of the most famous brands in the West is the 6666 of pioneer cattleman Burke Burnett. A primary reason for the fame of the brand is the widely known tale that young Burnett won his original ranch by drawing four sixes in a poker game, then designing a brand to celebrate his winning hand. But the poker game seems to have no basis in fact; as a young man Burke apparently purchased a hundred longhorns wearing the Four Sixes, and he bought the brand as well. His brother later used a reverse version of the brand, 9999.

Samuel Burke Burnett was born on the first day of 1849 in Missouri, moving to Texas with his family when he was seven. The Burnetts settled on Denton Creek in North Texas, where Burke's father, Jeremiah, raised livestock. After the Civil War, Jeremiah Burnett conducted roundups and became one of the first Texans to drive cattle to Kansas. In 1868, when he was nineteen, Burke rode as a cowboy for Wiley Robin on a trail drive to Abilene.

The next year Burke returned to Abilene at the head of ten drovers and 1,700 longhorns, in which his father had given him a financial interest. Despite losing a number of horses to Indian raiders en route, young Burnett completed the drive successfully. His share of the profits allowed him to begin buying and selling cattle of his own.

In 1873 he headed up the Chisholm Trail with 1,100 longhorns, but upon arrival in Wichita he learned that the bottom had dropped out of the market because of the Panic of 1873. Instead of selling at devastating prices, Burnett moved his herd into Indian Territory and grazed his animals on the Osage Reservation. In the fall of 1874 he made the short drive to Wichita and sold his herd for a profit of $10,000.

Burnett then bought 1,300 steers on the Rio Grande and Nueces River and drove them north to open range on the Big Wichita River about fifteen miles south of the future town of Wichita Falls. The cattle were fattened and sold, giving Burnett claim as the first Texas rancher to buy a herd of steers and fatten them for market. Indeed, the Waggoners and other contemporary ranchers quickly adopted this plan, which became common practice. In 1877 Burnett helped establish the organization that would

Burke Burnett built this 6666 barn in 1908 on his ranch near Guthrie. The barn has been moved to the Ranch Heritage Museum in Lubbock.

The main house of the 6666 near Guthrie, erected in 1917.

The oldest building erected by Burke Burnett, the Four Sixes commissary stands on the outskirts of tiny Guthrie, just south of the ranch's headquarters compound.

become the Southwestern Cattle Raisers Association, serving on the executive committee for forty-five years until his death.

Burnett had married in 1870, and he and his wife became parents of a son and daughter, Tom and Annie. Burnett hauled lumber and other materials from Fort Worth to the Big Wichita and built the first frame house in Wichita County. Recognizing that the days of open-range ranching were numbered, Burnett began buying land at twenty-five cents an acre. Within a few years he had 1,300 acres under cultivation, planting corn and oats for feeding, and he ranged 25,000 head of cattle and 2,000 horses and mules.

By the early 1880s, however, farmers and small ranchers were settling throughout the area, just as the growing Four Sixes herd needed more range. Looking across the Red River, Burnett became friendly with Quanah Parker, chief of the Comanches, and arranged to lease 300,000 acres of reservation land at six and a half cents an acre per year. Burnett also helped fellow cattle barons Charles Goodnight and W. T. Waggoner negotiate large leases with Quanah Parker for their respective operations, which brought welcome income to the Comanches. When the federal government later decreed that cattlemen vacate such leases so that these lands could be opened to homesteaders, Burnett brought Quanah Parker to Washington, D.C. In a personal interview with President Theodore Roosevelt (a former Dakota rancher), Burnett won a two-year delay, allowing cattlemen to make an orderly transition. In 1905 Burnett hosted Roosevelt in a memorable wolf hunt in Indian Territory.

Foreseeing the need to acquire title to more land, in 1900 Burnett purchased 141,000 acres and 15,000 cattle in King County from the Louisville Land and Cattle Company. This property, centered around the county seat at Guthrie, would become the headquarters of the Four Sixes, but other acquisitions would include 57,000 more acres in King County, 108,000 acres near Paducah to the north, 30,000 acres near Iowa Park in Wichita County, and 108,000 acres on Dixon Creek between Borger and Panhandle in the Texas Panhandle.

As early as the 1870s Burnett began improving his longhorns with Shorthorns, but like most Texas ranchers he soon switched to Herefords. By the 1890s Four Sixes cattle were "remarkable for smoothness of looks and evenness of weight," and Burnett's Herefords became regular prizewinners at livestock shows. Burnett

DEVILISH NOISY, BUT DEUCE HANDSOME

Raised on the Texas frontier, Burke Burnett received only two years of schooling. But his pioneer upbringing instilled in him courage, self-reliance, aggressiveness, and a strong work ethic. He was energetic and friendly, and even after becoming a multi-millionaire he exhibited an informality born of traditional frontier hospitality. Impressed after his first meeting with the noted cattleman, a young Englishman described Burnett as "Devilish noisy, but a deuce handsome chap."

Although gregarious by nature, Burnett was capable of lethal reaction to danger. When he was building his Four Sixes Ranch, the North Texas region was plagued by rustlers. Upon finding some of his steers re-marked with another brand, Burnett cut them out and drove then back to his range. Soon he was challenged by three hardcases, Jack King, Pat Wolford, and Jim Garrison. When King went for his gun, Burnett put a .45 slug in his head. Wolford and Garrison fled, and Burnett was quickly acquitted on grounds of self-defense.

In 1893 rustler Farley Sears, who previously had killed his own brother, stole calves belonging to Burnett and other ranchers. Sears was arrested, but he managed to gain his release. He threatened to kill Burnett and confronted the rancher in the men's room of a hotel in Paducah. But once again Burnett dropped his adversary with a fatal shot, then won acquittal through self-defense.

also enjoyed raising horses, mules, and hogs for market; he sometimes sold as many as 2,000 head of horses at one time.

Early in the twentieth century oil was discovered on his Wichita County land, then on the Dixon Creek ranch. The Wichita County community of Nestorville changed its name to Burkburnett in 1907, and the town sprouted a forest of oil derricks. Burnett shifted his ranching activities to the western properties, discontinuing the 6666 brand in Wichita County in 1910.

By this time Burnett's home was in Fort Worth. After the death of his first wife, Burnett married the widow of a physician in 1892, and to this union was born Samuel Burke Burnett, Jr. Burnett continued to visit his ranches regularly, but he became extremely active in the civic affairs of Fort Worth. He built the city's tallest office building, helped found the First National Bank, donated a city park, and, upon his death in 1922, willed a fortune to Texas Christian University.

Bud Arnett, ranch manager for the Louisville Land and Cattle Company, was hired in the same capacity by Burnett when he bought the spread, and Arnett provided continuity of leadership until 1930. His son-in-law, Horace Bryant, replaced Arnett as Four Sixes manager, but left the ranch after two years. George Humphreys, who hired on at the Four Sixes in 1918 as a teenaged bronc buster, assumed managerial duties in 1932, and did not retire until 1970, after fifty-two years of service to the Four Sixes. Such loyalty and longevity, along with an infusion of oil capital for ranch improvements, has maintained an extremely high quality of live-stock and infrastructure. With nearly half a million acres on four divisions in the hands of a great-granddaughter of Burke Burnett, the Four Sixes is the largest ranch in Texas under sole ownership — and that ownership directly descended from a remarkable pioneer cattleman.

Owner of the 6666 Ranch, Burke Burnett pioneered the steer-buying and feeding business. He and three generations of his family are inside this mausoleum.

⌒⊃ THE LONG S RANCH

The Long S was the West Texas domain of cattle baron Christopher Columbus Slaughter. C.C. Slaughter began handling cattle as a boy during the 1840s for his father, George Webb Slaughter, who established a family ranch in Palo Pinto County in 1856. C.C. (also called "Lum") stood off Indians during the Civil War, began driving cattle to markets after the war, and by 1870 established himself as the dominant figure in the Slaughter ranching enterprise. He was an expert judge of cattle and range conditions, as well as a shrewd trader, and he confidently engaged in banking, mercantile, and real estate interests. In 1873 C.C. moved to a fine home in Dallas to oversee his business ventures, but ranching in West Texas remained a predominant activity.

By 1878 Slaughter had formed his enormous Long S Ranch out of unclaimed lands only recently cleared of buffaloes and Comanche warriors. The Long S stretched along fifty miles of the Colorado River and varied its width from forty to eighty miles, a range of 2,400 square miles. Slaughter occupied the principal water sources within this arid area, and through the purchase and lease (state lands leased for as little as three cents per acre) of select acreage, the cattle baron for several years managed to control 1.5 million acres, which was referred to as "Slaughter country." Sprawling across five counties, the Long S was the second largest ranch in West Texas — only the vast XIT of the 1880s would surpass the Long S in size in the region.

That same year Slaughter placed the permanent ranch headquarters at German Springs, twenty miles north of Big Spring and near the recently completed Texas and Pacific Railroad. Slaughter organized his great ranch into four divisions: Indian Canyon on the north; Buffalo Division to the west; Sulphur Division along the southwest; and Rattlesnake Division to the east.

By 1882 there were 40,000 cattle grazing on the Long S, with an annual calf production numbering 12,000. Slaughter accompanied 350 three-year-old Durham-longhorn steers on the railroad to St. Louis, where he sold them for $27,000. The price Slaughter arranged, seven cents per pound, reputedly was the highest ever paid at that time for grass-fed beef, and thereafter Slaughter enjoyed a national reputation as the "Cattle King of Texas."

The roundups held each May and September were enormous affairs. One of the largest roundups ever conducted in Texas took place on the Long S in May 1884. A pasture half a mile square was readied, and roundup crews were sent out in different directions to camp overnight ten miles away. At dawn each crew spread out to contact the crew on either side, then the twenty-mile-wide circle began to contract. By 10:00 in the morning 15,000 cattle had been herded into the pasture, but the cutting and branding went on until sundown.

On such roundups Slaughter often had to employ as many as eighty riders. Slaughter was popular among his men, who enjoyed good working conditions and tended to stay on the Long S for many years. Jack Alley worked for Slaughter for four decades, eventually managing the Long S. Gus O'Keefe was the first manager of the Long S, providing capable leadership until he left to establish his own ranch in 1888.

Slaughter determined to enclose his range, beginning with a drift fence in 1883. At a cost of about $250 a mile (barbed wire sold for fifteen cents per pound, posts were ten cents apiece, and labor was ten cents a rod), within two years 100 miles of fence had been strung about the Long S. As early as 1883 Slaughter placed some of the first Texas windmills on his range. The windmills cost from $400 to $700 apiece, pumped from depths of sixty to eighty-four feet.

In January 1885 a severe blizzard killed from 5,000 to 10,000

A HALF-MILLION-DOLLAR RIDE

In 1881 Slaughter accepted an offer of $500,000 from two Englishmen for the Long S. Three days after the Englishmen left to take over the ranch, Slaughter learned that they had no money. Slaughter sent his ten-year-old son, Bob, on a legendary horseback ride to head off the transfer. Using three mounts, the boy raced 300 miles in forty-one hours to complete the famous "half-million-dollar ride." Thereafter Slaughter declined all offers for the Long S.

The Slaughter corral, rebuilt at the Ranch Heritage Museum in Lubbock. Rank horses were tied to the snubbing post, and the circular corral with no sharp corners protected broncs and riders.

Long S cattle, and drought and bitter storms the next winter continued to exact a heavy toll. While many cattlemen were driven out of business during this period, the optimistic Slaughter remained convinced that prices eventually would rise, and he traveled widely to find favorable sales for Long S beef. In 1886 Slaughter sold 10,000 head of stock cattle for a reported $140,000 to the newly established XIT, and by reducing his Long S herd by one-fourth he eased the strain on his drought-stricken range. The next year he hired a veterinarian to teach Long S riders how to properly spay cows. Spayed cows gain weight, providing Slaughter with fat animals to sell.

Slaughter then appointed his rakehell son, Bob, to run the Long S. A deeply religious man with strait-laced habits, Slaughter ordered Bob and two of his brothers to take the "Kealy cure" for alcoholism in Kansas City in 1894. Bob relished outdoor work, but his carousing and frequent inattention to the ranch caused his father to place Jack Alley in charge of the Long S in 1911.

Early in his career as a cattleman, Slaughter had crossbred Texas longhorns with Durham Shorthorns. Soon after establishing the Long S, however, he became a proponent of Hereford cattle. Although Herefords were ill-suited for trail drives, the proximity of railroads to the Long S during the 1880s offset this disadvantage. In 1884 Slaughter bought ten carloads of Hereford bulls and shipped them to the Long S, and in 1897 he paid $50,000 for a splendid herd of 1,900 Herefords (including sixty bulls) developed by Charles Goodnight on the JAs. That same year Slaughter visited Hereford farms in the North and Midwest, purchasing numerous fine bulls. He paid $2,500 for Ancient Briton, a British bull which had been award-ed the first prize at the 1893 Chicago World's Fair. Slaughter's bulls were pictured in trade periodicals, and his stationery soon contained a color etching of Ancient Briton and the proud boast that C.C. Slaughter owned the "Champion Hereford Herd of the World."

In 1899 Slaughter attended an auction in Kansas City, deter-mined to buy Sir Bredwell, the champion Hereford bull of the 1898 Omaha Exposition. Before a shouting, ecstatic crowd, Slaughter announced an opening bid of $1,000, then outbid the field for an unprecedented purchase price of $5,000. Sir Bredwell and Ancient Briton, however, were not destined for the Long S; along with other top-quality bulls, these animals were placed on the "Slaughter Hereford Home," a 2,000-acre alfalfa farm near Roswell, New Mexico, then on a new West Texas property termed the Lazy S.

For many years Slaughter inspected his western ranches twice annually, in addition to maintaining close control over his cattle operations by correspondence. Each June and October he boarded a Texas and Pacific train in Dallas and journeyed by rail to Big Spring. He then traversed the Long S and his other ranch properties by a horse-drawn coach (which for a time sported a painting of Ancient Briton on one side and Sir Bredwell on the other), before returning by rail to Dallas. In the spring Slaughter judged the con-dition of his cattle and various ranges, issued orders for ranch improvements, attended roundups, and determined whether to ship cattle to market. His fall routine included stops at watering sites, evaluations of winter pastures, and personal supervision of cattle selections for market. Slaughter continued to travel widely to find the best markets for his herds.

A relentless drought during the early 1890s caused the Long S to lose money for the first time in 1894, despite the sale of $100,000 worth of cattle that fall. Passage by the Texas Legislature in 1895 of the Four-Section Act, which permitted settlers to obtain title to 2,560-acre tracts for just $80 down, threatened the Long S and similar sprawling ranches. In 1897–98 harsh winter conditions, followed by a spring drought, killed 2,000 Long S cattle.

Slaughter fought to maintain his holdings, but by 1905 the Long S was reduced to 250,000 acres. Four years later he allowed a promoter to offer the remaining land for sale to farmers, but dry weather intervened and Slaughter kept 200,000 acres of the Long S. Bob Slaughter had permitted the fences, range, and watering locations to deteriorate, but when Jack Alley was placed in charge the Long S quickly was restored to a profit-making condition. More than two dozen wells with windmills were placed about the range, which soon supported more than 8,000 cattle, and in 1915 the Long S produced a net profit exceeding $106,000.

Despite declining health, Slaughter continued to visit the ranch each year, customarily spending the summer at Long S headquarters. Slaughter died in 1919 at the age of eighty-two, but contrary to his wishes his nine surviving children and wife split his property ten ways. Within two years of its founder's death, the Long S was divided into small tracts.

↶ THE LAZY S RANCH

As C.C. Slaughter rapidly improved his herds with expensive Hereford bulls late in the 1890s, he decided to increase his West Texas acreage with another large ranch. He intended to own outright title to this new spread, but farmers were swarming into West Texas to take advantage of the Four-Section Act of 1895 and of the windmill technology which made it possible to cultivate arid lands. The Texas Legislature and county governments had become pro-farmer, and a general public bias had developed against big cattlemen. But Slaughter enlisted Fount Oxsheer and other ranchers who were indebted to his Dallas bank, and in 1898 these agents quietly began acquiring large tracts on behalf of the well-known cattle baron.

By 1901 Slaughter owned 246,699 acres in Cochran and Hockley counties, for which he had paid $240,000, in addition to $95,000 to Oxsheer and other agents. Slaughter also obtained another 50,000 acres in leases, some from his own cowboys, who gained title through the Four-Section Act, then leased their 2,560 acres to the boss.

Originally flat-roofed, the stable is 75 feet long and is located on the west side of the compound.

Unique two-story dugout, built about 1890 as headquarters of the Lazy S, Whiteface Camp, and currently displayed at Lubbock's Ranch Heritage Museum.

The main house, 80 feet long and 15 feet wide, on the north side of the compound built in 1915 as Lazy S headquarters. The bunkhouse, which faces the main house, is similar.

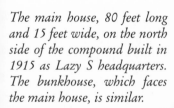

The milk house, erected just beside the main house.

In 1898 Slaughter hired as foreman Hiley T. Boyd, who had cowboyed in the area for five years. Then a crew was assigned to drill wells and erect windmills and dirt tanks. Each tank was supplied by two or three windmills; within a few years there were fifty-four windmills at twenty-four tanks, with an investment totaling nearly $60,000. Fencing crews enclosed four large pastures ruled by Ancient Briton, Sir Bredwell, and two other famous bulls. Indeed, at first Slaughter called his new ranch the Ancient Briton and Sir Bredwell divisions of the Long S, his big ranch located about forty miles to the southeast. But he began to view the new property as a separate entity, and after 1901 he referred to it as the Lazy S. The Long S brand ($\longrightarrow$) was shortened and placed on the animal's thigh rather than the side.

A reporter for the *Texas State and Farm Journal* accompanied Slaughter on a tour of the handsome new ranch. "The location is on a range which was never successfully developed until it came into Col. Slaughter's possession. Now it is one of the best-watered ranges in the country," admired H.W. Caylor. "This is truly the greatest purebred Hereford herd in the world," he added.

The Long S and Lazy S totaled a combined 1,373,000 acres, stocked with 54,500 cattle. For a time C.C. Slaughter was the largest individual taxpayer in Texas, as well as a noted philanthropist, especially to Southern Baptist institutions.

The first headquarters of the Lazy S was a half-dugout, but in 1915 Mexican craftsmen were brought in to build an adobe and concrete quadrangle in the style of a Spanish *hacienda*. The new headquarters was located about two miles south of Morton, the seat of Cochran County.

Four years later, C.C. Slaughter died. He had built the Lazy S as an "indivisible" legacy to his wife and nine children, but they soon divided all his properties into ten smaller ranches. The handsome *hacienda* complex continued to be utilized, and in 1937 oil was discovered on the old Lazy S. Today there are oil wells and irrigated fields on the "Slaughter Ranches," which still are headquartered near Morton.

⏚ T-ANCHOR RANCH

The T-Anchor in the Texas Panhandle was organized by shrewd, opportunistic land speculators. The T-Anchor ranch house was the first log cabin erected on the High Plains, and became the site of decisive action in the famous Cowboy Strike of 1883. Soon purchased by a British syndicate typical of the period, the T-Anchor later was broken up into farms and small ranches.

The ranch originated in the fall of 1877, when Leigh Dyer drove 400 head of cattle into the valley of Palo Duro Canyon (the previous year his brother-in-law, Charles Goodnight, had begun the famous JA Ranch in Palo Duro Canyon). Dyer and Arch Argo cut and hauled big cedar logs out of Palo Duro Canyon and built a dog-trot cabin, with an adobe smokehouse-blacksmith shop nearby.

But even as Dyer located his herd and erected a ranch head-quarters, two enterprising lawyers from Sherman were aggressively buying land certificates that would give them possession of Dyer's preempted spread. During Reconstruction the Texas state government issued certificates for unappropriated public lands in order to raise funds. By the late 1870s, a 640-acre land certificate could be picked up for as little as sixteen dollars, and Sherman attorneys Jot Gunter and W. B. Munson began borrowing money from friends and banks to buy as many certificates as they could find. While Gunter and Munson were hustling up certificates, they dispatched surveyor John Summerfield with a crew to the Panhandle, where the largest parcels of unappropriated Texas lands were located. "Others of us had the same opportunity," recalled surveyor W. S. Mabry, "but we did not have the vision; we did not believe that such a vast expanse of unoccupied land could be settled and become valuable in our lifetime."

Working on shares, Summerfield located and surveyed a variety of parcels for his partners, including a vast tract of Randall County that would become the T-Anchor Ranch. (Summerfield became overimaginative in drawing up his map of the region; he was indicted for attempting to defraud the Land Office, but Gunter managed to secure an acquittal when he was tried in Austin.) The partners bought Leigh Dyer's cabin and range rights.

In 1880 Gunter, Munson, and Summerfield purchased their

The T-Anchor headquarters cabin, built in 1877 from logs hauled out of Palo Duro Canyon, today stands outside the Panhandle-Plains Museum in Canyon. Note the outhouse beyond the dogtrot.

Log outbuildings at T-Anchor headquarters.

Rear view of the first house in the northernmost thirty-six counties of Texas.

first cattle, a herd from Louisiana, applying a GMS brand. Jule Gunter, Jot's nephew, bought Summerfield's interest in 1881, and the firm began to use a Crescent G brand. But when Jule bought a herd from Indian Territory branded with the T-Anchor, the other two brands were discarded and the ranch assumed the name of the new brand.

In 1881 and 1882 the T-Anchor put up the first big fence in the Panhandle, enclosing 240,000 acres of grass. The posts were placed eighty feet apart, so that when mustangs and antelope ran into the fence in full stride, the wire would give instead of break.

Jule Gunter was serving as ranch manager in the spring of 1883 when Panhandle cowboys attempted to organize and strike for higher wages. Several T-Anchor men quit the day before the April 1 strike deadline, and Gunter was tipped off that a delegation soon would ride to ranch headquarters to present demands.

Immediately preparing for trouble, Gunter forted up inside his cabin with his loyal men and a considerable arsenal. Assuming that any attack would come from the outlying smokehouse, a keg of gunpowder was placed inside the adobe structure and horseshoes and pieces of scrap iron were nailed to the walls. The fuse from the powder keg was several feet short of the cabin, but cook Gus Lee was assigned to crawl outside and light it. Only one delegate was sent to the T-Anchor, however, and he was met angrily by Gunter. He returned to strike headquarters at Tascosa with discouraging news of the T-Anchor fortifications. Gunter easily found replacements for the strikers, and so did other ranch managers, and the strike proved a complete failure.

Later in 1883 Jule and Jot Gunter sold out to W.B. Munson, who operated the T-Anchor for two more years. In 1885 Munson sold 225 sections and 24,000 head of cattle to the Cedar Valley Lands and Cattle Company of London. The English syndicate also leased additional lands, and at its peak the T-Anchor covered most of Randall and Deaf Smith counties, as well as parts of Armstrong, Briscoe, Castro, Oldham, and Swisher counties. Englishman John Hutson managed the T-Anchor from 1884 until 1902.

Texas land legislation in 1895 forced the T-Anchor to relinquish its leases to farmers and small ranchers. In 1902 the company decided, like most other large West Texas ranches, to divide its deeded lands into small blocks for sale to settlers. The old log head-quarters became the house of a farmer.

Near the T-Anchor headquarters had grown the town of Canyon, organized in 1889 as the seat of Randall County. In 1910 West Texas Normal (today West Texas A&M) was founded in Canyon, and in 1921 a history professor launched the Panhandle-Plains Historical Association, which later acquired the old T-Anchor cabin. Today the historic cabin is displayed on the university campus, and eighty acres of the former T-Anchor range north of Canyon has long been used by the agriculture department as an experimental station.

 ## RENDERBROOK-SPADE

Barbed wire revolutionized the western cattle industry. The famous fencing material was manufactured and promoted by Isaac L. Ellwood, who purchased a West Texas ranch that has been impressively operated by his family for more than a century. A former Forty-niner, Ellwood had settled into a mercantile and hardware business in DeKalb, Illinois, where his primary customers were farmers who needed improved fencing. In 1873 one of those farmers, Joseph L. Glidden, developed and patented the first practical barbed wire. Ellwood bought half-interest in Glidden's patent for $265, and within a few years a new brick factory and aggressive sales had begun to produce a multimillion-dollar fortune. In 1889 Ellwood, accompanied by his son, W.L., and longtime salesman, H.L. Sanborn, ventured into West Texas in search of a ranch in which to invest some of his surplus capital.

During the 1870s "a very large spring" was located south of present-day Colorado City by Capt. Joseph Rendlebrock and a Fort Concho cavalry patrol. By 1878 pioneer Taylor Barr had built a dugout at "Renderbrook Springs," a corruption of the captain's name. Within a few years Barr sold his dugout and open-range rights to D.H. and J.W. Snyder.

Experienced and successful cattlemen, the Snyder brothers built a two-room cabin with a dirt floor and thatched roof at the springs, then erected a larger residence and a bunkhouse of lumber, along with corrals and tanks. When the Ellwoods and H.L. Sanborn arrived in Colorado City in search of investment property, they were told about the Snyders' developing ranch. The Ellwoods hired a hack and headed south to inspect the 130,000-acre Renderbrook Ranch.

A purchase was quickly arranged, although the Snyder brothers retained ownership of their cattle. I.L. Ellwood headed back for DeKalb, but W. L. took a buckboard and driver from Renderbrook 200 miles north to the ranch of J.F. "Spade" Evans near Clarendon. Evans was nicknamed for the Spade brand on the left side of his cattle. Ellwood bought 800 head, then registered the brand in Mitchell, Coke, and Sterling counties, home to his new ranch, which commonly would be called "The Spades."

Within two years Renderbrook foreman W. L. Carpenter left the ranch, and the Snyder brothers suggested Dick Arnett as a replacement. The forty-four-year-old Arnett owned his own small ranch, but he left his oldest son in charge and moved his family to Renderbrook, where he would provide steady leadership for the next twenty-one years.

The Snyder brothers also suggested another ranch sale, a 128,000-acre property they had put together south of Renderbrook in Lamb and Hockley counties, with some spillover in Lubbock and Hale counties. The asking price was $2.50 an acre, which the Ellwoods met in part by shipping to the Snyders horses from their

One of the headquarters buildings at Renderbrook. (Author's collection)

DeKalb breeding opera-
tion. One carload also
went to Colorado City
to upgrade the Ellwood
ranch herd.

W. L. Ellwood had
accompanied his father
on horse-buying trips to
Europe since the late
1870s, but he began con-
centrating his livestock
trading and purchasing
on his West Texas ranch-

Headquarters of the South Spade was built in 1906.

es. Captivated by the open plains and by the mystique of ranching,
he spent as much time in Texas as in DeKalb. His wife preferred the
more cultivated lifestyle of the East, and made only occasional visits to the West.

The south ranch acquired in 1891 was dubbed the Spade, and
for the next decade and a half the Ellwoods continued to buy large
parcels. By 1906 the Spade had grown to 265,000 acres, and the
130,000-acre Renderbrook-Spade was eight to twelve miles wide
and ran nearly fifty miles north to south. The ranches were enclosed
with six-wire fences, while five-wire fences divided Renderbrook-
Spade into pastures averaging forty sections each. These enormous
pastures were connected by large wooden gates painted bright red,
so that they could be spotted by cowboys searching the long fence-
lines for passageways.

The Renderbrook was used primarily as a breeding ranch.
Young steers were transferred to the Spade for grazing until they
were sold as four-year-olds. The herd often numbered 30,000. For
two decades Renderbrook-Spade ran Shorthorns, but early in the
twentieth century the Ellwoods switched to Herefords.

I.L. Ellwood died in 1910 at the age of seventy-seven. His
daughters received cash for their portion of the enormous estate,
while W. L. and his brother, Perry, were left Renderbrook-Spade and
other properties. Perry Ellwood succeeded his father as president of
the First National Bank of DeKalb, and W. L. remained in charge of
the Spades.

By this time Dick Arnett functioned as general manager of the

CHUCK

One Renderbrook chuckwagon cook funneled a gallon can of apples into a cobbler only to find that there was no cinnamon. Resourcefully, he substituted mint-flavored Copenhagen snuff for flavoring.

Colonel Ellwood so enjoyed the chuckwagon fare of black cook Perry Bracey that he brought Bracey to headquarters each time the family visited Renderbrook. A creature of habit, Bracey shunned the headquarters cookhouse and worked off his chuckwagon parked behind the White House.

Bracey dubbed salt pork "Kansas City fish." Another black cook, Lee Proctor, boasted a larger culinary vocabulary when he grandly announced his "bill of fare" before each chuckwagon meal:

"hound ears" . . . . strips of sourdough fried in a Dutch oven
"Texas butter" . . . . flour gravy
"saddle blankets" . . . . . pancakes
"swampseed" . . . . rice
"clean up the kitchen" . leftovers
"Pecos strawberries" . . . beans
"pooch" . . . . cold biscuits and canned tomatoes
"lumpy Dick" . . . . boiled pudding
"spotted pup" . . . . rice, raisin, and cinnamon dessert
"French dish" . . . . same dessert with an exotic name "to make it more mysterious"

Spades for the munificent monthly wage of $208.33. He left the payroll in 1912 at the age of sixty-five, driving into retirement with a buggy and team given by the ranch for twenty-one years of expert leadership. Arnett was succeeded by a man who would nearly triple his own longevity of service to Renderbrook-Spade. Otto Jones was hired by Arnett as an eighteen-year-old cowboy in 1907. By 1910 Jones had earned promotion to wagonboss, and two years later he was elevated to manager at twenty-three.

As farmers moved onto the plains of West Texas, county governments increased property taxes rapidly in order to pay for the expected services. Facing ruinous taxes, during the 1920s the Ellwoods reluctantly offered portions of their land for sale at $35 per acre. Within a few years nearly 150,000 acres had been sold, reducing the ranch but producing needed revenue.

W. L. Ellwood died in 1933, and while his daughters loved the Spades, they had married and established homes in the East. Will Eisenberg, a longtime Ellwood employee, was transferred from the DeKalb offices to the Spades. Eisenberg and Otto Jones faced the problems of drought and depression, raising needed revenue by introducing sheep to the old cow outfit in 1935.

During World War II the Spades, like all big ranches, suffered manpower shortage. A few years after the war, 170,000 acres of New Mexico's historic Bell Ranch was added to the operation, in part by selling 30,000 acres of Renderbrook land to a farmland syndicate.

Drilling for petroleum had begun in 1929, but for years the only result was a shallow gas well which was piped to headquarters, providing gas for heating, cooking, and refrigeration long before the ranch enjoyed commercial electricity. Oil at last began to flow in the early 1950s, which helped the ranch suffer through the unprecedented drought that did not break until 1957. Only a skeleton herd — and crew — was maintained during the 1950s, and when the drought finally ended a deteriorated infrastructure had to be addressed before the ranch could be restocked.

"He offered me too much to turn down," mused Frank Northcutt after meeting with Otto Jones. "He was offering me a chance to improve a ranch. It was what I'd wanted to do all my life."

Jones wanted Northcutt to be assistant manager of Renderbrook-Spade with authority to make necessary changes. Old fences, corrals, tanks, and windmills were replaced, and Northcutt initiated a cross-fencing program to divide the enormous pastures into small enclosures of only a few sections, thus eliminating big roundups requiring large crews. Through the years the Renderbrook-Spade range had become infested with mesquite, cedar, catclaw and other brush species, and Northcutt launched a major brush control program.

Otto Jones retired in 1966 after six decades of service, and he was replaced as general manager of the Spades and the Bell property by Dub Waldrop, who held a Ph.D. from Texas A&M University. Combining practical experience with academic training, Waldrop modernized the tradition-bound old ranch. A rotation system was instituted, whereby three herds of cattle were rotated among four pastures, giving each pasture a rest in its turn. Waldrop also decided to switch from Herefords to crossbred cattle, traveling to government and university research stations, then to Canada and Europe to observe various breeds. Finally, the Spades bred Hereford, Angus, Swiss and Simmental in the first four-way cross of two British and two Swiss cattle breeds.

For more than a century Renderbrook-Spade has featured remarkable stability of operation and personnel, ownership by one family, and the flexibility and resources to pursue progressive developments. To a notable degree the history of the Spades is the history of ranching in West Texas.

TEXAS RANCHING
IN A CORRAL

Biggest Ranches: XIT (3,050,000 acres; 150,000 cattle)
JAs (1,335, 000 acres; 100,000 cattle)
King (1,150,000 acres; 75,000 cattle)

Best Cowtowns: Fort Worth and Tascosa

Most Popular Saloons: White Elephant, Waco Tap,
Maverick, and Red Light, Fort Worth
Equity Bar, Tascosa
Buckhorn, San Antonio

First Rodeo: Pecos — July 4, 1883

Best Black Cowboys: Bill Pickett and Bose Ickard

Most Famous Ranch Animals:
Old Blue, Goodnight Trail steer
Rosy Brown, XIT outlaw bronc "with wicked ways"
Ancient Briton   } Champion Hereford bulls of the
Sir Bredwell   Slaughter Ranches.

Most Violent Range War:
Pink Higgins vs. Horrell Brothers, Lampasas County.

R.I.P.   Pink Higgins, 1851–1913 — heart attack
Bill Horrell, b. 1838 — killed in Civil War
John Horrell, b. 1841 — shot dead ca. 1868
Sam Horrell, b. 1843–1930 — died in bed!
Mart Horrell, b. 1846 — lynched in 1878
Tom Horrell, b. 1850 — lynched in 1878
Ben Horrell, b. 1851 — killed in 1873 gunfight
Merritt Horrell, b. 1857 — killed by Pink, 1877

New Mexico

━━━ **JOHN CHISUM'S JINGLEBOB**

"I'm in great trouble because I cannot dispose of my stock as fast as it increases."

The rancher with this rather enviable "great trouble" was the "Cattle King of the Pecos," John Simpson Chisum. Sometimes confused with Jesse Chisholm, who blazed the famous Chisholm Trail, John Chisum moved from Texas onto the open range of Lincoln County and became New Mexico's first cattle baron.

Born in 1824, Chisum was reared on a Tennessee plantation and displayed such an affinity for cattle that he was nicknamed "Cow John." At the age of thirteen Cow John moved with his family to the fledgling community of Paris in northern Texas. Young Chisum helped his father on the farm, worked in a grocery store, dabbled in real estate, and was sworn in as county clerk in 1852. Within two years he had formed a ten-year partnership with Stephen K. Fowler, an eastern banker who wanted to invest in cattle ranching. Fowler advanced $6,000 to Chisum, who bought 1,200 head of cattle by the end of 1854, then located his first ranch north of Fort Worth.

During the Civil War, Chisum was designated a beef supplier for Confederate troops. He was paid $40 per head in Confederate

RANCHING IN NEW MEXICO

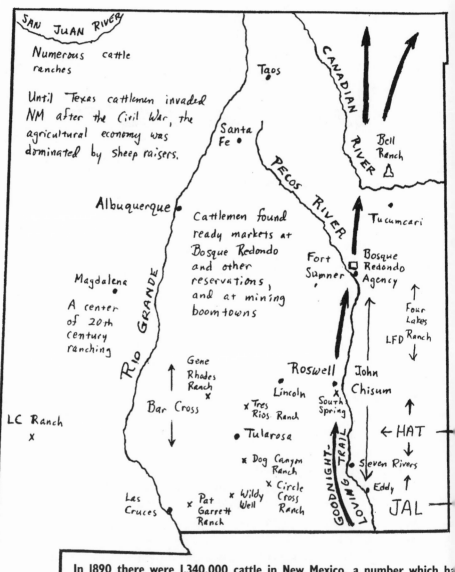

SAN JUAN RIVER

Numerous cattle ranches

Until Texas cattlemen invaded NM after the Civil War, the agricultural economy was dominated by sheep raisers.

Taos

CANADIAN RIVER

Santa Fe

Bell Ranch

Albuquerque

PECOS RIVER

Cattlemen found ready markets at Bosque Redondo and other reservations, and at mining boomtowns

Tucumcari

Fort Sumner

Bosque Redondo Agency

Magdalena

A center of 20th century ranching

RIO GRANDE

Four Lakes

LFD Ranch

Gene Rhodes Ranch x

Roswell

John Chisum

Bar Cross

Lincoln

x Tres Rios Ranch

South Spring

LC Ranch x

x Tularosa

← HAT →

x Dog Canyon Ranch

GOODNIGHT-LOVING TRAIL

Seven Rivers

x Circle Cross Ranch

Eddy

Las Cruces

x Pat Garrett Ranch

x Wildy Well

JAL

In 1890 there were 1,340,000 cattle in New Mexico, a number which ha[s] remained relatively constant through the twentieth century.

money, but following each sale he bought more cattle. By war's end he had converted the ultimately worthless currency to livestock. Indian raids and a desire for vast amounts of open range caused him to move his operation to West Texas during the war. In 1863 Chisum and a large crew drove 1,500 head of cattle to the confluence of the Concho and Colorado rivers in present-day Coleman County. Log huts and pens were built in a grove of pecan trees. The nearest neighboring spread was twenty-five miles to the east.

By this time Chisum had devised the "Long Rail" brand, a straight line burned on the left flank from hip to shoulder. The brand was run with the lower half of a metal rod bent into a half circle. The Long Rail was easy to alter with a running iron, but Chisum's cattle also wore the distinctive "jinglebob" earmark. A slice of a pocketknife left the bottom two-thirds of the ear dangling and swinging "just like an earring."

In August 1867 Chisum led a herd of his Jinglebobs to Bosque Grande, a New Mexico cattle exchange point of the Pecos River thirty-five miles north of present-day Roswell. He found a ready market at Fort Sumner, and by the next year Chisum had shaken hands on a partnership with Charles Goodnight, who had pioneered the Goodnight-Loving Trail across southwest Texas and northward into New Mexico along the Pecos. For the next few years Chisum would deliver as many cattle as possible to the Bosque Grande, where Goodnight's trail crews would drive the herds to sale points in Colorado, Kansas, and elsewhere. Thousands of cattle with varied brands were brought out of Texas, but if questioned about his title to the animals, Chisum would produce a lead pipe in which he kept papers giving him power of attorney to approximately 200 brands.

Chisum would put together a herd and drive it across the Goodnight-Loving Trail to the Texas line, where his younger brother and foreman, Pitzer Chisum, would take over with a crew while John returned to gather another herd. Also involved in the operation was another brother, widower James, who brought his children, Will, Walter and Sallie, to New Mexico. John remained a lifelong bachelor, although he was sociable and loved to joke and laugh. When one of his courtships ended, John Chisum simply chuckled: "The girl didn't court me enough."

In 1870 Chisum first visited the big square adobe house owned by A.M. Hudson at "South Spring," five miles south of Roswell. The "Square House" consisted of eight small rooms surrounding a twenty-foot square patio. A corral was attached to the only entrance, a roofed-over passage in the west wall. South Spring was an artesian stream which flowed five miles east to the Pecos. Chisum immediately recognized the well-watered fortification as a desirable base for his New Mexico ranch. James Patterson acquired the site from Hudson, but in December 1874 Chisum traded Patterson 2,400 head of cattle for the forty acres and improvements at South Spring.

Chisum's cattle multiplied rapidly. Repeated raids on his horse herd by Indians and rustlers would soon cause his "great trouble" — the inability to round up and drive his stock to market. By 1875 an estimated 80,000 Jinglebobs ranged from Anton Chico on the north to Seven Rivers on the south, and from the White Mountains on the west to Canadian on the east. The enormity of this range made Jinglebob livestock ready prey for thieves, and large numbers of cattle as well as horses were stolen. For a time Chisum hired rustler Jessie Evans to steal horses and cattle back from the Mescaleros who regularly raided his herds.

At the height of his operations Chisum employed at least 100 riders, many of whom were better gunmen than cowboys. Chisum often led his men in pursuit of stock thieves, and in April 1877 he and thirty "warriors" conducted a three-day siege of a nest of rustlers at Seven Rivers, a center for stock thieves on the Pecos above the Texas border. On another occasion Chisum and twelve of his men ran down a rustler who had helped steal three horses.

"We asked him no questions," related Chisum coldly. "Vegetation was scant there, but we took the highest we could find and dragged him up until his head was within two inches of the limb. . . ."

Although it was easy to alter the Long Rail brand, the Jinglebob mark was such a giveaway that thieves had to cut off the ears of Jinglebob cattle. On March 28, 1877, James Highsaw, a Chisum foreman, discovered a large number of recently removed ears at a public corral at Loving's Bend on the Pecos River. Highsaw confronted Dick Smith and shot the rustler to death. A score of Smith's friends pursued Highsaw to Chisum's headquarters, but they turned back at the sight of riflemen stationed atop the roof of the fortress-like Square House.

Influenced at least in part by incessant raids on his livestock, late in 1875 Chisum sold most of his cattle to Hunter, Evans and Company, a prominent St. Louis beef commission firm. Hunter and Evans agreed to pay Chisum $219,000, half down in cash, but it was understood that the cattle would be delivered piecemeal over a period of years. It was 1880 before Chisum completed his contract, delivering a total of 65,000 head of Long Rail cattle to various purchasing agents. By that time he had begun building a new herd. Since he had sold his famous Long Rail brand to Hunter and Evans, he adopted a U brand.

In 1877 Chisum entered a mercantile and banking enterprise in Lincoln, seat of vast Lincoln County, a lawless 27,000-square-mile region that was the largest county in the United States. Chisum's partners were attorney Alexander McSween and John Tunstall, a young rancher from England, and their firm was in direct competition with a similar operation dominated by L.G. Murphy, who led an anti-Chisum faction of small ranches and rustlers. Competition between the two firms was a major factor in the outbreak of the bloody Lincoln County War, in which Tunstall and McSween were murdered.

Two of the suspects in Tunstall's murder, Billy Morton and Frank Baker, were captured in March 1878 on the Pecos River below South Spring Ranch by a posse of "regulators" which included Billy the Kid. En route to Lincoln, this group stayed overnight at Chisum's ranch headquarters, and after departing the prisoners were gunned down. Later in the year Billy the Kid and four other Regulators were headed for a Fourth of July celebration at Chisum's ranch when nearly a score of hostile riders gave chase; a running fight ensued, and after the Kid and his friends reached headquarters there was intermittent sniping until nightfall.

As hostilities worsened, Chisum distanced himself from the Lincoln County War and tended to the pressing duties of running his ranch. He constantly patrolled his range, riding a roan called Old Steady or driving a buggy and team. Chisum carried binoculars and a revolver, with the holster buckled to the right side of his saddle horn. Although there were threats on his life, he usually traveled with at least one rider and rarely experienced dangerous situations. However, Chisum frequently had to defend himself from lawsuits over disputed livestock or rangeland. Late in 1877 and early in 1878, he was incarcerated in Las Vegas for eight weeks because he stubbornly refused to list his assets in response to suits filed against him.

Although the Lincoln County War climaxed with the five-day Battle of Lincoln in July 1878, rustlers (most notably Billy the Kid) continued to plague the area. By 1880, however, there was a concerted effort to crush the outlaws, and in 1881 Sheriff Pat Garrett killed the Kid.

Chisum finally had delivered the Hunter and Evans cattle, and as Lincoln County lawlessness declined he concentrated his efforts on developing a blooded Shorthorn herd and improving his South Spring headquarters.

THE CATTLE QUEEN OF NEW MEXICO

One of the most notable women of pioneer New Mexico, Sue McSween, established the Tres Rios Ranch and became known as the Cattle Queen of New Mexico —with material assistance from the Cattle King of the Pecos.

Born in Pennsylvania in 1845, Susan Hummer moved to Kansas and married attorney Alexander McSween in 1873. Within two years the couple moved to Lincoln, and as her husband became increasingly involved in the dangerous events leading to the Lincoln County War, Sue offered fearless support. When her home was besieged during the five-day Battle of Lincoln, Sue marched resolutely past opposing gunmen to appeal — in vain — for aid from Col. Nathan Dudley.

After her husband was slain she employed Huston Chapman as her lawyer, but he, too, was murdered. She tenaciously pressed her own suit, and acquired title to the ranch of the deceased Dick Brewer in payment of a note owed to her husband.

Sue remarried in 1880 but divorced in 1891. She sold the Brewer property and in 1883 purchased a ranch in the Three Rivers country on the west side of the Sierra Blanca Mountains. John Chisum, who had been in business with Alexander McSween, donated a small herd of cattle to help launch the Tres Rios Ranch. Sue erected a rock house and eventually built her herd up to 8,000 head. The Cattle Queen of New Mexico finally sold Tres Rios to Albert B. Fall in 1917 and moved to White Oaks, where she died in 1931 at the age of eighty-five.

When Chisum set cottonwood fence posts, the posts took root and grew into lines of great trees, a few of which still stand.

During the late 1870s, Chisum dredged two large irrigation ditches north from South Spring River: the Texas Ditch was pointed north from a distance east of the Square House, while the other, called Pumpkin Row, was begun about a mile west of the ranch corral. Ox wagons were sent to the Davis Mountains in Texas for cottonwood and willow saplings to line the ditches. Irrigation permitted Chisum to cultivate fields of alfalfa, millet, wheat, and other grains.

In 1881 Chisum spent $12,000 to build the "Long House" about 300 yards southwest of the Square House. Facing west, the adobe

In 1881 John Chisum spent $12,000 to build the "Long House."
— Courtesy Historical Society for Southeast New Mexico, Roswell

structure was sixteen feet wide and nearly 150 feet long. There were eight rooms, four on each side of a ten-foot-wide hallway. Chisum's office-bedroom, just north of the hallway, boasted a writing desk, a dictionary on a stand, a small safe, and a bedstead. The shingled, pitched roof covered verandas with plank floors which ran the length of the west and east sides. The "drinking canal," fed by an artesian stream, ran beneath the hallway east into the garden. The old Square House was razed, but some of the materials were used to build two flat-roofed, sixteen-by-twenty buildings: a commissary behind the northeast corner of the Long House, and, behind the southeast corner, living quarters for the black housekeeper, Aunt Mary Blythe. A large barn stood 200 yards to the northeast.

By this time competing cattle companies had established themselves along the Pecos, and Chisum's range now stretched "only" sixty miles along the river, from Salt Creek, above Roswell, south to Artesia. But Chisum continued to buy fine bulls, building his herd back to about 30,000 head. Unfortunately, Chisum's health began to decline, and increasingly he turned ranching operations over to James Chisum and Will Robert, who had married Sallie Chisum in 1880. James and Will began borrowing money to invest in the

ranch, but John's younger brother, Pitzer, was uncomfortable with this style of management. John requested a $100,000 settlement for Pitzer, who moved back to Paris, Texas, and married early in 1884.

Aside from an 1877 bout of smallpox which left his face pitted, John Chisum had enjoyed good health — until a large tumor developed on his neck beneath his right ear. New Mexico physicians could do little, and in July 1884 the

When John Chisum died in 1884, he was buried in a small family cemetery at Paris, Texas.

ailing Chisum, visibly moved at leaving his ranch, sought treatment in the East. He underwent surgery in Kansas City, then sought further relief from the mineral baths at a resort in Eureka Springs, Arkansas. But his condition worsened, and he died in Eureka Springs on December 22, 1884, at the age of sixty.

John Chisum left an estate valued at $500,000 to Will Chisum, Sallie Chisum Robert, and Walter Chisum. His heirs were unable to hold onto the ranch, and turned over the property in 1890 to M.J. Farris for $100,000. Will and his wife, Lina, were the last Chisums to reside in the Long House, where their daughter Josephine was born in 1889. After a fire damaged the rambling old house, it was torn down.

Railroad owner J.J. Hagerman bought the South Spring Ranch in 1892, then began working to finance construction of a line into Roswell. The railroad connection was completed in 1898, and Hagerman moved to the area two years later. A short distance northwest of the site of the Long House, Hagerman erected a three-story brick residence dubbed "The Manor." The third floor was a ballroom. Four other large buildings also were constructed, bringing the total cost of the new headquarters complex to approximately $50,000. But Hagerman had just begun to spend. He added land until he owned 6,746 acres, and he invested in numerous large-scale dam and irrigation projects. By the time he died in 1909, he had invested $2.5 million in the region.

Portions of the ranch were mortgaged by the heirs to try to maintain solvency, but in 1932 a court order of default and judgment awarded the property to Cornell University, which held the mortgage. In 1941 Cornell sold the South Spring Ranch to Howard E. Babcock, Sr., and later in the year Mr. and Mrs. Bert Aston bought the Hagerman mansion and a portion of the ranch from Babcock. Bats were living in the top of the old house, so the Astons leveled it off to one story. Robert Anderson purchased the historic property in 1968 and made it part of his Diamond A Cattle Company. Anderson sold out in 1993, and today the nucleus of Chisum's nineteenth-century empire is a 700-acre dairy farm. Remains of the dams and ditches built by Chisum are still visible, and Hagerman's turn-of-the-century structures stand near a historical marker.

In 1901 J.J. Hagerman built a three-story ranch house a few hundred yards northwest of John Chisum's long adobe, which was razed.
— Courtesy of Morgan Nelson

With the top story of Hagerman's 1901 house infested by bats, new owners Mr. and Mrs. Bert Aston took off two floors.

J. Phelp White, nephew of George Littlefield, came to Roswell in the 1880s to run his uncle's cattle operation on part of John Chisum's old range. Eventually White became the wealthiest man in the Pecos Valley, and today his home serves Roswell as a museum.

DOG CANYON RANCH

"Well, Oliver," said Perry Altman to his younger half-brother, Oliver Lee, "this country is so damn sorry I think we can stay here a long time and never be bothered by anybody else."

Dog Canyon and the surrounding countryside indeed looked barren and forbidding, and Lee would end up staying a very long time, only to be "bothered" by a number of people. Of course, Lee bothered — and buried — quite a few people himself, en route to becoming one of New Mexico's most successful cattlemen and politicians.

Cañon del Perro (Dog Canyon) is one of many deep canyons which gash the western side of the rugged Sacramento Mountains in southern New Mexico. The canyon walls rise abruptly 2,000 feet from the arid floor of the Tularosa Basin. Although the region receives less than ten inches of precipitation annually, a spring-fed stream runs year-round through the canyon and supports a lush greenery. The availability of water, along with the defensibility of this natural fortress, made Dog Canyon a favorite Indian campsite.

Apache raiders often escaped eastward through Dog Canyon, or set ambushes for pursuers.

Apache raids had just ended when, in the mid-1880s, Francois Jean Rochas squatted at the mouth of Dog Canyon. Born in 1843 in France, Rochas was a bachelor with a deep yearning for solitude and an expressed desire to be called Frank. Frank Rochas built a small rock house, along with stone fences and corrals. He tended a growing herd of cattle and cultivated a garden and an

Looking east into Dog Canyon.

orchard with apple, peach, pear, and cherry trees. The water which produced his abundance was a precious commodity, and he would have to protect it. In 1886 he was wounded by a bushwacker at his cabin. Rochas shot his attacker, but he continued to receive threats. On December 26, 1894, three men rode to his cabin and fatally wounded the fifty-one-year-old Rochas.

A year earlier, Oliver Lee ran an irrigation ditch a mile and a quarter to his new ranch headquarters on the flats southwest of Dog Canyon. Lee and Perry Altman came from their West Texas home in 1884 to investigate prospects in New Mexico, and they were strongly attracted to the west side of the Sacramento Mountains. They returned to Texas to bring their families and live-stock to New Mexico. In 1885 Perry and his bride settled on a claim eight miles west of Tularosa, while Oliver and his mother, twice-widowed Mary Altman Lee, started a little spread seven miles west of La Luz. Widely known as a crack shot, Lee first fired his guns in anger during an 1888 feud with a neighboring rancher, John Good. Good's son, Walter, was suspected of the murder of Lee's boyhood friend, George McDonald. Lee procured the bullet which had killed McDonald and wore it on a watch chain, and he was one of four men charged with Walter Good's death later in the year.

Ruins of the cabin at the mouth of Dog Canyon where Frenchy Rochas was killed in 1894.

Lee was released from custody and in 1889, at twenty-four, moved to the flats west of Dog Canyon. He put up the first windmill between Alamogordo and El Paso, which became known as Lee's Well. Soon Lee shifted to a point about a mile south of Dog Canyon. He began building an adobe ranch house, and in 1893 he ditched water in from Dog Canyon to his homestead. Also in 1893, Lee and Bill McNew trailed a herd of stolen cattle to a point near El Paso, whereupon Lee galloped toward rustlers Charley Rhodius and Matt Coffeldt and shot both men to death.

Although not generally regarded as a culprit in the murder of Frank Rochas, Lee had constructed extensive irrigation facilities from Dog Canyon to water his cattle and the large orchard and garden at his ranch headquarters. He expanded his operations as far as Wildy Well, a ranch thirty-two miles south of Alamogordo. Lee helped found Alamogordo, north of his Dog Canyon Ranch, in 1897. By that time Lee was a suspect in the murder of Col. Albert Jennings Fountain and his nine-year-old son, who had disappeared in the White Sands in February 1896. Famed manhunter Pat Garrett was appointed sheriff of Dona Ana County and placed in charge of the investigation. Bill McNew was arrested, but Lee eluded capture. Garrett and a posse finally jumped Lee and fellow suspect James Gilliland at Wildy Well on July 13, 1898, but Deputy Kent Kearney was fatally wounded and the lawmen retreated as the victorious fugitives rode for safety.

Lee slipped into Texas for several months, marrying Winnie Rhode in San Antonio late in 1898. Perhaps anxious to begin family life free of legal pressure, Lee returned to New Mexico early in 1899, surrendered to authorities, and was acquitted following a sensational trial in Hillsboro. Lee enclosed his adobe ranch house, built a slaughterhouse nearby, as well as an adobe barn and numerous outbuildings, and continued to improve his irrigation system. A reservoir west of the house contained goldfish to keep down algae. A schoolroom was added to the house, and a

Circle Cross ranch house.

PASÓ POR AQUÍ

Eugene Manlove Rhodes was born in Nebraska in 1869. His father was named agent of the Mescalero Apaches at Fort Stanton, New Mexico, where the family moved in 1881. Later they homesteaded a hardscrabble spread at an opening in the San Andreas Mountains that would become known as Rhodes Pass. Gene began cowboying at thirteen, working for the Bar Cross and other big outfits in the Tularosa country.

In 1892 Gene homesteaded a claim near his father's old place at Rhodes Pass and built a picket house. Pugnacious and a frequent participant in altercations, Rhodes sympathetically granted refuge at his lovely ranch to Black Jack Ketchum, Bill Doolin, and lesser outlaws. Rhodes was a friend of Oliver Lee, offering him haven following the murders of A.J. Fountain and his little boy, then helping to arrange his safe surrender to custody prior to trial.

While working on an irrigation pipeline for Lee in 1906, Rhodes quarreled with a man over a crap game and broke six beer bottles over his head. He promptly headed for New York, not to return to New Mexico for two decades. Rhodes had married a widow in 1899, and May encouraged Gene's urge to write. He eventually produced numerous western novels and short stories based on his experiences and acquaintances.

Mr. and Mrs. Rhodes finally returned to New Mexico in 1926, living for a time in Alamogordo across the street from Oliver Lee. Failing health prompted Gene to move to California, but he died in 1934. He was buried atop Rhodes Pass, with his epitaph the title of his classic short story, *"Pasó por Aquí."*

teacher was brought out from Alamogordo for the Lee children (there eventually would be nine little Lees). In 1907 James R. Fennimore and three other men tried to fence off Dog Canyon, but Lee, as courageous and contentious as ever, rode single-handedly to stop them and nicked Fennimore twice during an ensuing rifle duel.

Lee was approached in 1914 by James G. McNary and several other El Paso businessmen who proposed the organization of an

Oliver Lee began building his adobe ranch house near Dog Canyon (visible in background) in 1893.

The adobe barn falling into ruin, about 1940. The side at left originally served as bunkhouse.

Remains of the barn today, from the same view.

Oliver Lee's 1895 Winchester and revolver holster, on display at the Dog Canyon visitor center.

Later in his life, Oliver Lee made his home at 1200 Ohio in Alamogordo. For a time author Eugene Manlove Rhodes lived across the street.

enormous ranch with Lee as manager. They bought out Lee, then organized the Circle Cross Ranch with a capital of $800,000. McNary was president and Lee was vice-president and general manager. Headquarters, high in the Sacramento Mountains, became a showplace, featuring a big house, gardens, and a deer park. Numerous ranches were acquired, and the Circle Cross soon encompassed a million acres. Financial difficulties during the 1920s would cause the breakup of the vast spread.

A handsome, impressive man fond of good clothes, Lee became heavily involved in public affairs after the turn of the century. The prominent cattleman was repeatedly elected to the state legislature, either as a representative or senator, and he served as an officer or director of numerous business organizations. But he always packed a holstered .45 beneath his Prince Albert coat, even during sessions of the legislature. In 1919 Lee was driving in an automobile toward El Paso when his companion, Johnny Hutchings, was fatally wounded by F.M. Scanaland. It was rumored that the bullet was meant for Lee. The next year Scanaland was beaten to death under mysterious circumstances.

Although Lee had a controversial and deadly reputation, he inspired great respect and was warmly regarded by a large circle of friends. He retired to Alamogordo, where he died of a stroke in 1941 at the age of seventy-six. Lee's Dog Canyon Ranch and water rights earlier had been acquired by J.L. Lawson, who rented out the old house for years. The National Park Service acquired the headquarters area a year before Lee died. Today there is a visitor center and campground at Dog Canyon, as well as the ruins of Frank Rochas' stone house, and Oliver Lee's sprawling ranch house has been handsomely restored.

THE BELL RANCH

About four miles north of the Bell Ranch headquarters is the bell-shaped mountain that gave the big spread its brand and name. For more than a century and a quarter the Bell Ranch has maintained one of New Mexico's largest cattle operations.

In 1824 Don Pablo Montoya, a prominent landholder and stockraiser, obtained a 655,468-acre grant along the Canadian River in northeastern New Mexico. Montaya died in 1842 without developing this excellent grassland where Comanches and Kiowas rode freely. Other undeveloped land grants nearby included the smaller Pedro Armendariz Grant and Baca Number Two, one of five grants which had been allotted to Luis Maria Cabeza de Baca. In 1863, at a bend of the Canadian on Baca Number Two, Fort Bascom was built, but the army abandoned the adobe outpost in 1870.

Also in 1870, these grants became the subject of acquisition by Wilson Waddingham, a bold mining promoter, financier, and land speculator. Born in Canada in 1833, Waddy joined the California Gold Rush, then began to pursue various opportunities across the West and Mexico. Aggressively seeking heirs of the New Mexico grants, Waddingham acquired title to more than 754,000 acres by late 1872. His range manager, Mike Slattery, established ranch headquarters in the deserted adobe structures of Fort Bascom, located in the southeastern corner of the spread. Slattery kept the livestock close to the old fort, because the absence of troops in the region encouraged Indian raiders. The worst depredation, however, was committed by white outlaws, who killed five riders and carried off a large number of Waddy's horses in 1874.

Waddy liked to bring investors to the ranch, and after 1876 he tried to build a brick house to the north of La Cinta Creek. When the house burned before it could be completed, a new headquarters complex grew up nearby, centered around a sprawling adobe residence that would become known after whitewashing as the "White House." In the White House on January 7, 1883, Waddy hosted a luxurious seven-course dinner that may have been the finest meal served in New Mexico Territory to that date.

During the late 1870s, Waddy began upgrading his herd by annually acquiring from 100 to 500 Shorthorn bulls to breed with

his native range cows. Waddingham organized capital stock corpo-
rations — always with himself as president — to raise money to
purchase cattle. By the early 1880s the improved quality of the Bell
brand animals grazing in the Canadian country was clearly evident.
Further improvement occurred about 1886, when Mike Slattery put
a barbed wire fence around the Bell range, making it possible to
keep desirable cattle in and undesirables out. The fence stretched
140 miles around, enclosing several thousand acres of public
domain as well as land to which Waddy held title. In 1891 Slattery
reported that 52,000 cattle grazed the 791,000 acres of the Bell
Ranch, and that he expected to brand 17,000 calves in the spring.
Beginning in the mid-1880s, the Bell Ranch put up 400-500 tons of
prairie hay each year.

By 1890 Waddingham was in serious financial trouble. For
years he had poured most of his energies and resources into mining
ventures, real estate deals, and various other projects in other states.
Although most of these investments went sour, Waddy bought a
40,000-acre ranch west of Las Vegas, New Mexico, for his daughter
and son-in-law. He also tried, without success, to develop intensive
irrigation and colonization by farmers along the Canadian River.
His continual purchases of blooded bulls and the skilled management of Mike Slattery had produced a cattle herd widely noted for
its outstanding quality. From the mid-1880s on, however, overuse
of the range, drought, bitter winters, and depressed cattle markets
made the Bell Ranch consistently unprofitable. Apparently, Waddy
put up the ranch as security for loans he could not repay, and by
January 1891 the property had been taken over by the New York
City firm of Brown and Wells, attorneys and financiers. Wilson
Waddingham continued to finance speculative schemes on credit
until he died suddenly in 1899.

Willard Brown became president of the Bell Ranch Irrigation
and Cattle Company. Soon afterwards a flash flood washed away a
large earthen dam under construction on the Canadian, and irrigation activities were halted. Slattery worked for the new owners until
1893, when the ranch was acquired by John Greenbough and James
Brown Potter, two leading New York City financiers. (Wilson
Waddingham was involved with the new owners, but was unable to
regain personal control of the ranch.) Slattery was replaced as ranch
manager by Arthur Tisdall, a big, friendly Irishman who had en-

tered the cattle business in Texas a decade earlier. Tisdall operated his own West Texas spread for a time, then he was hired by the widow of John Adair to manage the famous JAs at an annual salary of $3,000. When offered $4,000 by Greenbough and Potter, Tisdall accepted the challenge of rebuilding the Bell Ranch, physically as well as financially.

The buildings and equipment of the bankrupt ranch had fallen into disrepair, and long stretches of fence were down. New Mexico Territory still was lawless: shortly after Tisdall assumed his duties, fence rider Mike Lantry was murdered; range manager Baldy Haynes was shot to death by an irate cowboy in his room at headquarters; thieves broke into the headquarters commissary; and rustlers constantly raided the vast Bell range. Furthermore, there were no more than 12,000 cattle on the ranch. Greenbough and Potter had been led to expect more than twice that total. Still, Tisdall repaired fences and structures, and he began to purchase Hereford bulls with the intention of rebuilding the herd to 25,000 improved cattle.

Tisdall patiently initiated the peaceful removal of numerous Bell Ranch trespassers. Injunctions were obtained against and payments were made to Mexican squatters who lived in five small communities within the ranch boundaries. Their livestock grazed off Bell Ranch grass, and so did sheep whose owners regularly placed their flocks within the fences. Injunctions also were obtained against these interlopers.

Relief from Tisdall's pressing duties came when he met Francis Harriott, a lovely Scotswoman touring the West with her mother. A whirlwind courtship resulted in a wedding on February 24, 1896, and Mrs. Tisdall soon expanded and improved the White House. Two years later, Tisdall was stricken with pneumonia and died on April 8, 1898; his wife was too distraught to attend the funeral.

Another Irishman, Charles M. O'Donel, was hired to replace Tisdall. O'Donel was a graduate of Sandhurst, Britain's military academy, and he had served as an officer during the Zulu War. Captain O'Donel had resigned his commission, moved to the Texas Panhandle, in 1885, and soon began working on a ranch. After serving as a ranch manager in the Panhandle, he was employed by the Bell Ranch. His arrival to take on-site control was delayed by the lengthy illness of his wife in Clarendon, Texas. O'Donel tried to handle his managerial responsibilities by correspondence, but he finally moved to the ranch after his wife died in September 1898.

O'Donel's military back-
ground refined an austere de-
meanor, a strong sense of organization, and meticulous attention to detail. Influenced by the
ideas of Arthur Tisdall, John
Greenbough directed his new
manager to divide the Bell
range into fenced pastures for
better herd management, and
"to conduct the ranch entirely
as a breeding ranch and have
nothing on it but she stock,
with the exception of the
calves, which would be sold as
yearlings." For more than three
decades O'Donel would imple-

*Irishman Charles M. O'Donel, man-
ager of the Bell Ranch, 1898-1932.*
— Courtesy Panhandle-Plains
Museum, Canyon

ment Tisdall's vision of the Bell Ranch as a cow-and-calf operation.
In addition to his fencing activities, O'Donel installed windmills
and dug stock tanks, bringing water to numerous grazing areas
across the ranch. (As a general rule, cattle lose weight if they must
walk more than two and a half miles to water, and the ideal Bell herd
of 25,000 head required about 375,000 gallons daily.)

In 1899 the Red River Valley Company, based in New Haven,
Connecticut, and headed by bank president E.G. Stoddard,
acquired the ranch. Stoddard firmly supported the policies of
Greenbough and O'Donel; he traveled west to visit the ranch each
autumn — and was killed by a fall from a horse in 1923.

O'Donel remarried in 1908, to Louise Harral of New Orleans.
Louise gave birth to two daughters, Betty and Nuala, and as the
girls grew older the family acquired a two-story brick house in
Denver to better facilitate their education. Mother and daughters
lived there during the school months, while O'Donel visited
Denver as often as possible. The girls spent long summers at the
ranch, learning to ride and rope. They made good hands and eagerly joined roundup and branding activities. Each evening, however,
they doffed their work clothes and donned proper dresses for dinner in the White House.

Early in his tenure O'Donel decided to upgrade the cattle by purchasing Shorthorn and Hereford bulls in approximately equal numbers, and soon he developed a herd of registered Shorthorns. Although isolated during its early existence, the Bell Ranch enjoyed access to nearby railroads by the turn of the century. In 1902 the Rock Island built into Tucumcari, four miles south of the south boundary fence and thirty-three miles south of headquarters. In the same year O'Donel granted permission for the Rock Island to build a branch line northwestward across Bell range to the coal-mining town of Dawson. The ranch enjoyed great savings on reduced shipping costs, including those on the large windmills O'Donel began to erect (sixty-eight were in operation by 1915). Also during this period, telephone lines were installed, allowing O'Donel to talk with cattle buyers, eastern directors, attorneys in Santa Fe, merchants or bankers in Tucumcari, and countless other business contacts and friends. As at many another ranch, the first wires were strung atop existing fence posts, and the improved efficiency was incalculable.

Cattle prices were at an all-time high during World War I, but markets dropped sharply at war's end, and the 1920s brought an extended drought. Conditions worsened, of course, during the Great Depression.

Since 1915 O'Donel had suffered increasing pain, caused by a fall from a horse, and he sometimes was confined to bed for a week at a time. In 1932 he finally retired, staying on at headquarters among the 3,000 books he had collected during nearly four decades as ranch manager. O'Donel died at the White House on December 20, 1933, at the age of seventy-three.

The new manager was thirty-eight-year-old Albert K. Mitchell, who assumed his new duties on January 1, 1933. A graduate of Cornell University with a degree in animal husbandry, Mitchell returned to his native New Mexico to manage the family ranch near Clayton. His father, T.E. Mitchell, in 1896 started a quality Hereford herd, and Albert would energetically implement a program of selective Hereford breeding on the Bell Ranch. Early in his tenure Mitchell struggled with deflated Depression markets and with an unprecedented drought in 1934 and 1935. As the Bell range burned beneath a moistureless sky, Mitchell shipped most of his herd to rented or leased pastures and feedlots, then eventually sold the cattle directly from the pastures and feedlots as markets improved. In

LONGEVITY

Tradition is an integral part of cattle ranching. At the historic Bell Ranch, traditions were established and perpetuated by men who served the spread for a great manay years. The pattern was set by managers who provided stability of leadership for long periods: Michael Slattery (ca. 1870–1894); Charles M. O'Donel (1898–1932); Albert K. Mitchell (1933–1947); George Ellis (1947–1970); and Don Hofman (1970–1986). Current manager Rusty Tinnin has worked for the Bell Ranch since 1973.

Mark Wood first drew Bell wages in 1919, and by the time the ranch was divided in 1947 he had served as wagon boss for twenty-three years. In charge of Mark's remuda for more than two decades was Ralph Bonds, who worked for the Bell Ranch a total of forty-eight years. Benito Encinias ran the "chain gang" (maintenance crew) for twenty-eight years, beginning late in the nineteenth century. Seferina Estrada was the longtime cook at the White House, while "Judge" Naylor worked as ranch blacksmith for many years.

The longest tenure of all was sustained by Moises Romero, who went to work for the Bell Ranch haying crew as a boy during the 1880s. By the 1890s he was drawing pay as a cowboy, and he continued as a Bell rider even after acquiring his own spread nearby. An expert with horses and cattle, Romero knew the trails, waterholes, and creek crossings on the Bell. Moises died after becoming ill during the fall work in 1946, having become a legend through six decades of service on the Bell Ranch.

1934 his wife died, leaving three children, and six months later his father suffered a fatal heart attack. His mother reared the children on the family ranch, although Albert's namesake would die of leukemia at the age of eight.

A workaholic who traveled constantly on business, Mitchell persuaded the directors in 1937 to purchase an airplane and hire a pilot. The next year the Republican Party persuaded him to run for governor, and he conducted a whirlwind campaign by plane. Mitchell lost, but then he tried unsuccessfully to win a U.S. Senate seat in 1940. It was an era dominated by Democratic politicians, and Mitchell did not again seek elective office, although he was extremely active in stockmen's associations.

Under Mitchell's management, the Bell Ranch began to return dividends to its shareholders in the late 1930s, and market conditions were extremely favorable during World War II. But by the end of the war it was widely rumored that the ranch was for sale; the most influential directors remembered the post-World War I disasters; Mitchell was considering retirement to his ranch; rising property taxes struck hard at an enormous block of land like the Bell Ranch; escalating paperwork required by federal and state governments added an additional burden; and the sale of more than 238,000 acres in 1918 brought the realization that an impressive amount of money could be produced by selling the big spread. In 1946 the Bell Ranch was placed on the market for $3 million.

A purchase was arranged the next year by Albert Mitchell and two partners, who intended to sell the ranch and split the profits equally. The 469,185-acre property was divided into six ranches, ranging in size from the 39,023 acres of the "Old Bell Farm" to the 130,855 acres of the "Old Bell Headquarters." Mrs. Harriett Keeney of Connecticut bought the headquarters spread and kept the Bell brand, and her ranch manager, George Ellis, lived in the White House with his wife. In 1970, the same year that the headquarters buildings were placed on the National Register of Historic Places, the Keeneys sold out to William Lane, who intended to piece together as much of the old ranch as possible. By the time Lane was killed in an automobile accident eight years later, the Bell Ranch comprised over 300,000 acres of the original range. Don Hofman served the Lanes as manager until his retirement in 1986, when he was succeeded by Rusty Tinnin. The Tinnins make their home in the White House, where Waddingham, Slattery, Tisdall, and O'Donel once lived, and Bell cattle still amble to water down the red banks of the Canadian and graze in the shadow of the Bell Mountains.

NEW MEXICO
RANCHING IN A CORRAL

Biggest Ranches:
   John Chisum (150 miles long; 100 cowboys; 80,000 cattle)
   Hat Ranch (150 miles long x 35 miles wide, from West Texas to Pecos River; in best years 60 cowboys branded over 30,000 calves)
   Four Lakes or "Windmill" Ranch (1,500,000 acres; 45,000 cattle)
   JAL Ranch (60 miles x 60 miles in SE corner of NM; 40,000 cattle)
   LC Ranch (Lyons-Campbell range was 60 miles long and 40 miles wide in 1880s, with headquarters at Gila)
   Bar Cross (100 miles north to south, San Marcial to Dona Ana; Rio Grande on west to San Andreas Mountains on east)

Best Cattle Towns:
   Phoenix, near Eddy (Carlsbad); Roswell; Lincoln

Best Rustler's Town: Seven Rivers

First Rodeo: Santa Fe "roundup" — June 10, 1847

Best Black Cowboy:
   Frank "Chisum," former slave who accompanied his former master to New Mexico and became a crack Jinglebob hand.

Most Violent Range War:
   Lincoln County War, originated in part because cattle baron John Chisum suspected L. G. Murphy of rustling; triggered by the murder of rancher John Tunstall.

R.I.P.   Billy the Kid, and a host of others.

3

Oklahoma

♫ DOG IRON RANCH

The birthplace of Oklahoma's favorite son, Will Rogers, was one of the territory's most prominent pioneer ranches. Will's father, Clement Vann Rogers, was born in 1839 on his father's small ranch in the Going Snake District of the Cherokee Nation, Indian Territory. Clem's father died when he was three, and when his mother remarried two years later the little boy obstinately refused to attend the ceremony, then threw rocks at the departing newlyweds. When Clem left the ranch at seventeen, his mother and stepfather gave him twenty-five cows, a bull, four horses, and two slaves that had belonged to his father.

Even more important than this impressive stake was the fact that Clem Rogers was five-sixteenth Cherokee. The land of the Cherokee Nation was held in common for use by any of its citizens, free of charge. Clem was ambitious, intelligent, hard-working, and gruff, and he began ranching in a grassy, well-watered frontier region known to the Cherokees as Cooweescoowee. In 1858 he married a fellow nineteen-year-old, Mary America Schrimsher, who was a quarter-blood Cherokee (their eight children would be nine-thirty-second Cherokee).

In 1861 the Cherokee Nation allied with the Confederacy, and Clem Rogers enlisted in the Cherokee Mounted Rifle Regiment. By

113

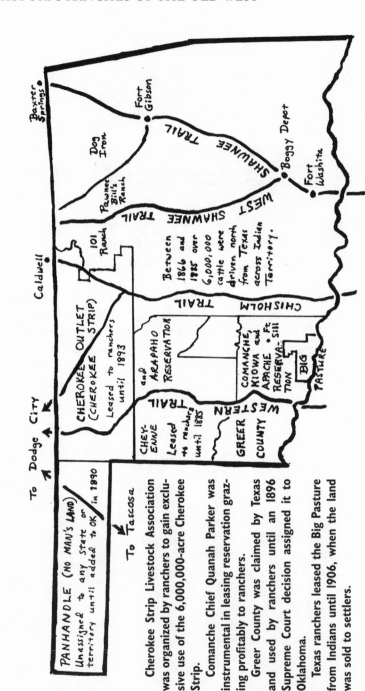

Ranching in Oklahoma

To Dodge City

To Tascosa

PANHANDLE (NO MAN'S LAND)
Unassigned to any State or
territory until added to OK in 1890

CHEROKEE OUTLET
(CHEROKEE STRIP)
Leased to ranchers
until 1893

Caldwell

Baxter Springs

Fort Gibson

Dog Iron

Pawnee Bill's Ranch

SHAWNEE TRAIL

WEST SHAWNEE TRAIL

Boggy Depot

Fort Washita

101 Ranch

Between 1866 and 1885 over 6,000,000 cattle were driven north from Texas across Indian Territory.

CHEYENNE and ARAPAHO RESERVATION
Leased to ranchers until 1885

CHISHOLM TRAIL

COMANCHE, KIOWA and APACHE RESERVATION

Ft. Sill

BIG PASTURE

GREER COUNTY

WESTERN TRAIL

Cherokee Strip Livestock Association was organized by ranchers to gain exclusive use of the 6,000,000-acre Cherokee Strip.

Comanche Chief Quanah Parker was instrumental in leasing reservation grazing profitably to ranchers.

Greer County was claimed by Texas and used by ranchers until an 1896 Supreme Court decision assigned it to Oklahoma.

Texas ranchers leased the Big Pasture from Indians until 1906, when the land was sold to settlers.

the end of the war he was a captain, but his growing family had moved away for safety, and federal troops had overrun his ranch, freed his slaves, and confiscated his horses and cattle.

For three years Clem worked as a farmer and teamster to save enough money for another start. In 1868 he began a second ranch, about seven miles east of his original spread in the Verdigris valley of the Cooweescoowee country. A Cherokee named Tom Boot had built a log cabin on the site, but Clem paid him $25 for the sixteen-foot-square dwelling. Although Maj. Dewitt Clinton Lipe helped Clem buy his first herd of steers, Rogers soon bought out his partner. Clem's approach to ranching was to purchase herds of Texas longhorns each spring, fatten them for a year on his bluestem rangelands, then ship them to market in St. Louis.

His business strategy proved immediately profitable. Clem ran cattle on a V-shaped range bounded by the Verdigris and Caney rivers. Each side of the V was about twelve miles long, and helped form two sides of a natural corral. The top, near the Kansas border, was fifteen miles long, but Clem's cowboys regularly rode that line to keep his cattle contained. He used about 60,000 acres, but he still paid no rent or taxes. Texas cattle could be purchased for only a few dollars a head, and Clem's only other significant expense was cowboy wages. Every year he marketed 2,000 to 4,000 steers with his CV brand, and sale prices usually were several times his costs. He also began to grow wheat in his bottomlands, acquiring the sobriquet "The Wheat King."

Clem was the driving force behind his ranch. Cowboy Ed Sunday later expressed admiration for his former boss: "Take it from me, Clem Rogers was a ranchman any way you wanted to look at him. . . . He was a keen trader, knew cattle, was one of the best riders I have ever known, and, get this, he never carried on his cattle business in an easy chair. He was a hard worker and if he ran one horse down he'd get another. He took part in the roundups, branding and shipping of steers, and, believe me, he was boss of the range." Indeed, Clem commanded universal respect; at the age of twenty-three he was elected a delegate to the 1862 Cherokee Confederate Convention, and for the remainder of his life he would be pressed into public service.

By 1875 Clem had completed a substantial home, less than a mile from the Verdigris. Built of hewn logs, the seven-room, two-

Completed in 1875, the Rogers ranch house was built of hewn logs that were plastered inside and weatherboarded outside. Will Rogers was born in his parents' bedroom, to the right of the doorway.

story house was weatherboarded, painted white, and comfortably furnished. The parlor piano was the only one in the region, and the Rogers home became a center of social activity. The ranch also boasted a sturdy barn, two other dwellings, and half a dozen other structures.

Each spring Clem conducted a calf roundup, taking six to eight weeks to brand his new calves. In the summers there was a beef roundup, then several leisurely drives to the nearest railhead, about forty miles to the north to Coffeyville, Kansas. Each fall a general roundup gathered all of the livestock on the Rogers range; strays from other herds were taken to their owners, while Clem's cattle were driven to winter range.

In 1889 the Missouri Pacific Railroad laid track which virtually bisected Clem's range, and which created the town of Oolagah just six miles southwest of the Rogers ranch house. At least now a shipping point nearby adjoined his ranch, and Clem opportunistically brought in 3,500 cattle from Texas. Two years later Clem built the first barbed-wire fence in the area, an eight-mile line from the Verdigris River to the railroad. Then the Missouri Pacific fenced in its right-of-way, forcing Clem to pasture a reduced herd east of the railroad, while expanding his farming activities on the west. With a fenced-in range, Clem introduced Shorthorn bulls to the region, then later added Herefords.

Clem and Mary Rogers lost three of seven children in infancy, but on November 4, 1879, when they were forty, their last child was born — William Penn Adair Rogers. When "Willie" was three his only surviving brother, Robert, died of typhoid fever. Now the only son as well as the baby of the family, Willie was indulged generously by his parents. Given his own ponies at an early age, he was attended by a cowboy responsible for his safety. A black cowboy, Uncle Dan Walker, introduced the little boy to rope tricks. Willie roped incessantly, and was never happier than on horseback.

In 1890, when Willie was still just ten years old, his mother died of amoebic dysentery. During the next three years Willie's sisters married and moved away from home, and his father wed Mary Bible, who was half his age.

Shortly after Mary Rogers died — and at about the same time he was organizing a fruitless pursuit of the Dalton brothers, who had stolen horses from the ranch — Clem received a large shipment of Texas cattle. There were seventy-five dogies, and he gave the motherless animals to his motherless little boy. Willie devised a "Dog Iron" brand, shaped like the fireplace andirons at the ranch house. One day his home place would be called the Dog Iron Ranch.

Willie was a restless youth, and he proved troublesome at a series of schools which his father forced him to attend. When he was eighteen he left

Will Rogers, the most famous product of the Dog Iron Ranch.
— Courtesy Will Rogers Memorial, Claremore, Oklahoma

Kemper Military Academy, where he had piled up a record 150 demerits, and headed for the Texas Panhandle, where he cowboyed for several months. During that same period, the Curtis Act of 1898 put an end to the Rogers Ranch. The Curtis Act abolished tribal laws and required allotment of Cherokee lands to individuals, which would open vast tracts to white farmers. As Cherokee citizens, Clem and Will received a meager allotment totaling 148.77 acres. Clem managed to acquire a few adjoining allotments, but his 60,000-acre ranch now was reduced to a farm. Clem sold his livestock and rented his land to a farmer from Illinois, then moved to Claremore, where he helped to found a bank.

Willie returned from Texas early in 1899, and Clem persuaded him to run his Dog Iron livestock on what was left of the old ranch. Willie was dismayed that the ranch house now was vacant of furniture, but he and a cowboy companion, Spi Trent, enthusiastically decided to establish an old-fashioned cow camp. They built a twelve-by-twelve log cabin, only to discover that they had forgotten to put in windows. Their furniture consisted of a double bed, two boxes which served as chairs, and a cookstove, on which they perpetually kept a pot of beans.

The Dog Iron herd required little work, but Willie became bored and increasingly spent his time with girls or at cowboy contests. After a year and a half he drifted away, eventually becoming a show business star and a national treasure. When Clem died in 1911 Will inherited the remnant of the ranch, which later passed to his heirs.

A quarter century after Will's tragic death in 1935, his children donated the property to the State of Oklahoma, with the understanding that the house would be moved to high ground. The Oolagah Dam and Reservoir were under construction, which would flood the home site. The house was moved upward from its original location to a site overlooking the lake. Recently, Amish builders were engaged to construct a barn in the fashion of Clem's original barn, and today the Dog Iron Ranch is a state park preserving and enhancing the Will Rogers birthplace.

This reconstruction of Clem Rogers' original barn was recently built by Amish workers, who use construction methods and materials of the 1870s.

Interior view of the Rogers barn reconstruction.

Water trough and pump in front of the Rogers ranch house.

101   101 Ranch

"This is the most interesting corner of the United States," reported noted writer Will Irwin. Irwin and his wife had come to the 101 for a one-day visit, but had succumbed to the charms of Oklahoma's most famous ranch and stayed a week.

The Irwins were among the parade of celebrities who enjoyed the fabled hospitality of the 101 Ranch. Teddy Roosevelt, Warren G. Harding, Randolph Hearst, Harry Sinclair, William Jennings Bryan, and John D. Rockefeller, Jr., headlined the host of political, business, and civic leaders who trooped to the 101. Gen. John J. Pershing and Adm. Richard Byrd sampled some of the same rustic pleasures, as did heavyweight champs Jack Dempsey, Gene Tunney, Jack Johnson, and Jess Willard. Will Rogers warbled cowboy songs deep into the night, and John Philip Sousa was initiated into the Ponca Indian tribe. Irresistibly drawn to the 101 were western movie stars William S. Hart, Tom Mix, and Hoot Gibson, along with Wild West Show empresarios Buffalo Bill Cody, Pawnee Bill, and Col. Zack Mulhall.

The National Editorial Association held its annual meeting at the 101 in 1905, and returned twenty years later. The National Realtors met at the 101, and so did the American Association of Petroleum Engineers and the Oklahoma Sunday School Convention and the Cherokee Strip Cowpunchers Association, as well as myriad other organizations. In 1934 a journalist reminisced in the *Daily Oklahoman:* "It was one continuous entertainment of guests, social, political, business leaders, writers, explorers, actors, the prominent men and women in every line."

This remarkable ranch was founded by pioneer cattleman George Miller. Born in 1841, he was reared on his grandfather's Kentucky plantation, where he nurtured an affinity for livestock. Miller married in 1866, and within two years the couple had a son, Joe. By 1870 Miller sold his share in the plantation in hopes of building a western cattle empire, and with his wife and son headed for California. Stopping for the winter in Newtonia, Missouri, he traded with area settlers for hogs, which he converted into hams and bacon, and he set out for Texas with ten wagonloads of hog meat. At fifty pounds of meat per steer, he traded for four hundred head of cattle. Driving his cattle up the old Shawnee Trail, he per-

A CORNY VERDICT

In 1872 George Miller rode to San Saba, Texas, to buy cattle, and he established his crew in a camp in Bosque County. Miller unwisely showed one of his men where he had hidden $2,000 in gold. When Miller went to retrieve the money two days later, it had disappeared. The crew proceeded on to San Saba, picked up the purchased herd, then camped at the Bosque County site on the way back to Oklahoma.

"The next morning we had a trial," related Jim Rainwater, Miller's longtime foreman. "The hats of all the boys were placed on the ground near a sack of shelled corn." One by one, each cowboy placed a grain of corn in the hat of the man regarded as the thief.

"The result was that all the corn was put in one man's hat — we all suspected the same fellow," said Rainwater. "This man picked up his hat, looked at the corn, shook it out, put on his hat, got on his horse, and rode away, and I presume he is riding yet."

suaded the Quapaw Indians to let him pasture his herd on their reservation in northeastern Oklahoma, a few miles south of Baxter Springs, Kansas.

Recognizing his opportunity to build a cattle empire, Miller continued to buy cattle in Texas, graze them on his reservation pastures, and market them in Baxter Springs. Miller built a home in Newtonia, where his wife, Molly, gave birth to a daughter, Alma, in 1875, and to another son, Zachary Taylor, in 1878. The next year, with his herd now numbering in the thousands, Miller leased from the government two large pastures in the Cherokee Strip totaling 60,000 acres.

The Salt Creek pasture was centered about twenty miles south of Hunnewell, Kansas, and a few miles further south was the Salt Fork pasture. Headquarters of Miller's new ranch went up on the Salt Fork River near present-day Lamont. There was a three-room log cabin with a dirt roof, a log corn crib, a horse barn and corral, and a branding pen and chute. In 1880 Miller moved his wife and children from Newtonia to Baxter Springs, where a third son, George, was born the following year.

In 1881 Miller adopted the brand his sons eventually would make famous. He bought out a small ranch near the Salt Fork called the Bar-O-Bar, with a —O— brand on the horns. A 101 mark was easier to read, and easier to read still when applied to the left hip instead of the horns. (Zack Miller always insisted that the brand originated with the 101 Saloon in San Antonio.) The 101 cattle continued to multiply on Cherokee Strip grass until 1892, when the government announced that the next year the Strip would be opened to homesteaders.

Resourcefully, Miller looked eastward, beyond the boundary of the Cherokee Strip, to lands belonging to the Ponca tribe. By the time of the spectacular land rush of 1893, Miller had moved his herd twenty miles east to a 50,000-acre Ponca lease along the Salt Fork River, at an annual fee of $32,500. A headquarters dugout was erected on the south bank of the Salt Fork; the front was built of planks, the rear was burrowed into the side, and grass grew on the sod roof. A family home was constructed in Winfield, Kansas, about thirty-five miles north of the new ranch. By this time Joe Miller worked as a partner with his father, while Zack and George were learning the business.

In 1903 "Colonel" Miller (although reminiscent of his Kentucky background, the title was purely deferential) began building a showplace ranch house on the north bank of the river across from the dugout. However, he contracted pneumonia in April of that year and died at the age of sixty-one. Five thousand mourners gathered on the lawn of Miller's unfinished prairie palace. The family proceeded with construction of the columned, three-story "White House," and on October 31 Alma was married there to attorney William Henry England, who later would handle legal affairs for the ranch. The White House was not yet completed when the wedding was held, but the entire clan gathered for Christmas at their new home.

Molly Miller received $30,000 from her husband's life insurance and Alma was given a separate ranch property as settlement, while the three brothers were instructed to keep the 101 intact through a trust. Fence had been put up around the 50,000 acres of leased land, but Molly used her $30,000 to buy 3,720 acres from various Poncas. Indeed, until her death in 1918 at the age of seventy-two, she remained involved in the business of the 101. Eventually,

Who Won?

During the fall of 1888, while the presidential campaign between incumbent Grover Cleveland and Republican challenger Benjamin Harrison was heating up, George Miller decided to offer a large number of his aging horses for sale to farmers. At the sale Miller, realizing that most of the farmers were Republicans, instructed the auctioneer to announce that the horses could be purchased with notes at ten percent interest — payable when Grover Cleveland was reelected.

The farmers felt challenged, and loudly yelled out bids considerably higher than normal. The best of the horses was worth only $75, but most bids exceeded $100. To the delight — and apparent profit — of the farmers, Cleveland won the popular vote but Harrison took an electoral majority. Miller prudently filed the notes in his rolltop desk, however, and four years later retrieved them when Cleveland defeated Harrison in the rematch of 1892. Miller turned the notes over to an attorney to collect at forty percent accrued interest, and the majority of the notes were reluctantly paid in full.

the Millers purchased 17,492 acres of deeded land. Along with about 10,000 acres of leased land with preferential rights and other leased lands which varied in amounts from year to year, at one point the 101 reached a maximum size of 110,000 acres.

Joe, Zack, and George ran the 101 in a brotherly arrangement unusual for its long-term amicability (Joe left the ranch after a 1917 disagreement, but a reconciliation was effected within two years). None of the brothers received a salary, but all shared the profits and drew upon the ranch bank account freely and with the complete trust of the others. The oldest brother, "Colonel Joe" (he was named an honorary colonel by the governor of Oklahoma in 1915) had functioned as ranch manager for his father and continued in that capacity. "Colonel Zack" (his honorary colonelcy was conferred by the governor in 1923) handled livestock sales, purchases, and trades with the consummate skills of a natural-born horse trader. The youngest brother (although his honorary colonelcy was awarded in 1919, he was always called "George L") was a gifted financier who increased the family fortune through investments and who devised an accounting system noted for its thorough simplicity.

All three brothers, with their families and mother, lived in the White House, but in 1909 the home burned to the ground. Although house and contents were insured for only $7,500, the family immediately built a new, $35,000 White House near the original site. Construction was of concrete and steel, along with an asbestos roof; only the floors, doors, and ornamental woodwork were susceptible to flames. The first floor was dominated by a vast living room and dining room; there were nine large bedrooms, each with a separate bath, on the second floor; and the third floor beneath the roof was a large open area, with billiard tables and enough four-poster beds to sleep one hundred people. Practicing the legendary hospitality of Kentucky colonels and western cattle barons, the Millers encouraged the constant presence of guests. Colonel Joe relished welcoming visitors on his enormous front porch with a universal, "Come on in, children."

In addition to a cattle herd that sometimes numbered as many as 25,000 head, the Millers ran a large Holstein dairy operation, raised thousands of hogs, and maintained hundreds of horses and mules. At least 15,000 acres of former range land were brought under cultivation, much of it for livestock feed but with wheat as the leading cash crop. Immense quantities of vegetables were harvested, and extensive fruit and nut orchards were planted. In 1910 a natural gas well was brought into production on land owned by the 101, and within a few years oil wells were pumping on the ranch. A refinery was built near headquarters, and surplus gasoline was sold from the ranch filling station. A $30,000 dairy/creamery was built, which included an ice cream plant. Also constructed was a meat-packing plant capable of dressing 500 head of cattle and 1,000 hogs per month. There was a two-story cannery and cider works, as well as an ice factory and a power plant. A general store, hotel for employees, trading posts for the Poncas, tannery, commercial harness shop, blacksmith shop, and garage also were part of the complex, not to mention numerous houses, barns, and corrals.

Perhaps the most impressive structure, aside from the White House itself, was a 12,000-seat rodeo arena, "the largest and finest in the Southwest." In 1905, when the National Editorial Association first came to the ranch, the Miller brothers staged a combination rodeo and Wild West show with 101 cowboys and Ponca Indians. Zack Miller went to Texas to hire Bill Pickett, already famous as the "Dusky Demon," to perform his bulldogging act, and the "first cowgirl," Lucille Mulhall, was brought in to demonstrate her championship roping and steer-tying skills. The feature attraction was famed Apache war leader Geronimo, who killed his last buffalo with a bow and arrow — but from an automobile driven alongside the shaggy beast.

Newspapers, of course, publicized the extravaganza widely, and a phenomenal crowd of nearly 65,000 gathered to watch the performers and eat buffalo barbeque and ice cream. Enthusiastic over this overwhelming response and the welcome publicity, Joe eagerly arranged to put his show on the road. The 101 Wild West Show departed in the spring of 1906 on two special trains; 126 performers boarded Pullman cars, while animals and equipment were loaded onto 126 freight cars. Appearing in Madison Square Garden in New York, the Jamestown Tricentennial Exposition in Virginia, Convention Hall in Kansas City, and the Chicago Coliseum, the show was a resounding success.

The Miller brothers built a big rodeo arena on the ranch, staging performances while honing the acts. Star of the 101 was Bill Pickett, who controlled wild steers by biting their lower lips, the way he had seen cow dogs operate when he was a boy in Texas. Future movie stars Tom Mix, Buck Jones, and Hoot Gibson toured with the 101, along with many other spectacular performers. The 101 traveled to Canada and Mexico City and, in 1914, to England. When war erupted in 1914 the British government impounded the horses and vehicles of the 101 "for public service" for $80,000 compensation, but the show was quickly reorganized. The 101 Ranch became Oklahoma's most famous institution, and the guest book at the White House bristled with noted names.

After World War I, however, Wild West shows found conditions increasingly difficult. Western movies became extremely popular, and rodeos enjoyed a growing following. While audiences for Wild West shows thus found other entertainment, expenses for large traveling troupes became burdensome. The show began to lose money. Then, in 1927, Joe died of carbon monoxide poisoning

THE DUSKY DEMON

Born in Texas in 1871, Willie M. Pickett grew up fascinated by working cowboys. Many cattlemen used bulldogs to work their livestock; the dogs controlled steers by biting their upper lips. When Pickett was ten he saw a bulldog handle a longhorn in that fashion, and a few days later the boy tried it himself. He seized a calf by the ears, sank his teeth into the animal's upper lip, and, to the astonishment of onlookers, threw it with ease.

As a teenaged cowboy, when he was in thick mesquite and brush that frustrated his attempt to build a loop, Pickett would lean from his horse, grip the steer's long horns, and "bulldog" it. Pickett always approached a steer from the left side, the side from which his horses were accustomed to being mounted and dismounted.

By the time he was sixteen he began to show up at fairs and similar events to demonstrate his unique method of wrestling steers. Delighted by the applause of appreciative crowds, Pickett began to travel extensively for exhibitions.

In 1903 he became affiliated with Dave McClure, a gifted promoter who booked him for appearances all across the West. Pickett was a mixture of African, Cherokee and Caucasian heritages, and McClure billed him as the "Dusky Demon" — a mild subterfuge, since blacks were barred from entering most competitions. As the only professional "bulldogger" in the world, the Dusky Demon proved to be a great attraction.

Employed by Joe Miller in 1905, Pickett made the 101 his home for most of the rest of his life. Pickett and his wife, Maggie, eventually produced nine children. Their seven daughters lived to adulthood, but both sons died in infancy.

For years Pickett was the only headliner of the Miller Brothers 101 Ranch Real Wild West Show, and the sole performer listed in the billing. He continued to enter rodeo competitions into his fifties, dogging an estimated 5,000 animals.

While cutting out horses on the 101 in 1932, a freak accident toppled Pickett from his mount, and a kick from an outlaw bronc crushed his head. Forty years after his death, the Dusky Demon, the only individual performer known to have originated a rodeo event, became the first black man elected to the National Rodeo Hall of Fame.

while working on a car in a closed garage at the ranch. Thousands attended the funeral services conducted on the massive front porch of the White House; troupers in costume served as ushers while Joe's Arabian steed, Pedro, stood in front of the casket wearing a $10,000 saddle.

Early in 1929 George's car skidded on an icy road and overturned, breaking the driver's neck. Once more thousands paid their respects at a 101 funeral, and George was buried beside his mother and brothers in the ranch cemetery.

Within a year and a half the 101 lost its longtime ranch manager and its financial officer. The 101 had been heavily mortgaged to keep the show afloat and to finance oil drilling. Zack continued to take the show on tour, but the agricultural operations which had provided a solid financial base for all other 101 activities began to show signs of neglect, then suffered staggering blows with the onset of the Great Depression in 1929.

By 1931 the 101 was in receivership, and the following year everything except the White House and its furnishings went before an auctioneer. A few days after the auction the Miller Brothers' greatest star, sixty-year-old Bill Pickett, was helping cut out horses that had been sold when his head was crushed by a kick from a bronc. Zack resolutely tried to put a 101 show on the road again, but his efforts continually proved unprofitable, and in 1936 he lost the White House to creditors. Zack continued to stage exhibitions until he died in a Waco hospital in 1952.

The federal government bought the White House in 1937 and launched a resettlement program on 8,000 acres of former 101 land. Tenants were moved onto small tracts and later allowed to purchase their farms. The Farm Security Administration sold the deteriorating buildings of the old 101, including a dilapidated White House. The White House was razed and the materials were hauled away, and today little remains to denote the once-magnificent ranch complex except a historical marker on the west side of Highway 156.

PAWNEE BILL'S BUFFALO RANCH

The 101 began as an open-range cattle ranch, and even after the Miller Brothers' Real Wild West Show focused attention on the entertainment branch of the operation, the ranch remained an enormously productive and diversified agricultural empire. About twenty-five miles southeast of 101 headquarters, however, a picturesque little ranch took shape early in the twentieth century as the showplace home of one of the greatest of the Wild West show empresarios.

Born in 1860, Gordon William Lillie worked as a trapper and a cowboy before becoming a teacher at the Pawnee Agency in Indian Territory. He became fluent in the Pawnee language and was used as an interpreter for Pawnees who participated in Buffalo Bill Cody's Wild West show. After learning the ropes, "Pawnee Bill" organized his own show in 1888. The next year he led a group of boomers into Oklahoma, newly opened to settlers, and built a log cabin on a claim atop Blue Hawk Peak. Blue Hawk was a Pawnee chief, and the peak was near the future site of Pawnee.

Pawnee Bill spent the next couple of decades touring with his popular Wild West show. He stayed at the ranch during the off-seasons, building his spread to 2,000 acres and raising a herd of buffalo. The outfit became known as Pawnee Bill's Buffalo Ranch. Pawnee Bill's wife, May Lillie, starred for many years as a horseback marksman, then retired to the log cabin at Blue Hawk Peak.

By 1909 Pawnee Bill and May were planning a splendid ranch house and outbuildings. The next year's tour brought him $200,000 in profits, half of which went into a magnificent two-story residence on Blue Hawk Peak. The building stone was quarried from Blue Hawk Peak, and the paintings, hangings, and furniture created a handsome western motif. In addition to the buffalo herd, horses roamed the ranch and Pawnees pitched their tepees on the property.

Pawnee Bill and his wife delighted in hosting guests and conducting tours of their charming ranch. Nearby he established Old Town and Indian Trading Post for tourists, but the facility burned in 1939. May died after an automobile wreck in 1936, and Pawnee Bill passed away at Buffalo Ranch just before his eighty-second birthday in 1942. Today the ranch is maintained by the State of Oklahoma and is open to the public.

For years Pawnee Bill and his wife stayed in a three-room log cabin at their ranch.

The top of a stone gate pillar proclaims the entrance to "Pawnee Bill's Buffalo Ranch."

Pawnee Bill's magnificent ranch house was built in 1910 at a cost of $100,000.

Gateway to the Buffalo Ranch atop Blue Hawk Peak.

The vast barn was built in 1926.

OKLAHOMA RANCHING IN A CORRAL

Most Famous Cattle Trails:
Chisholm Trail and Western Trail

Most Famous Movie Stars of the Future:
Will Rogers, Dog Iron Ranch
Tom Mix, 101 Ranch
Hoot Gibson, 101 Ranch
Buck Jones, 101 Ranch
Mabel Normand, 101 Ranch

Most Famous Black Performers:
The "Dusky Demon," Bill Pickett, 101 Ranch

Most Famous Female Performer:
Lucille Mulhall, who became a top hand on her father's Logan County ranch. The "first cowgirl" was world lady champion in steer roping and broke records in competition with men.

Most Notorious Cowboy-Turned-Outlaw:
Bill Doolin, who cowboyed on several I.T. ranches from 1881 until becoming a fugitive after an 1891 shootout; then led a gang of thieves before being killed in 1896 by a posse in Lawson, Oklahoma.

Best Chuckwagon Cook:
Oscar Brewster, Cherokee Strip ranches

Most Famous Cowboys Organization:
Cherokee Strip Cowpunchers Association, organized 1920

Best Cowboy Museum: National Cowboy Hall of Fame and Western Heritage Center, Oklahoma City

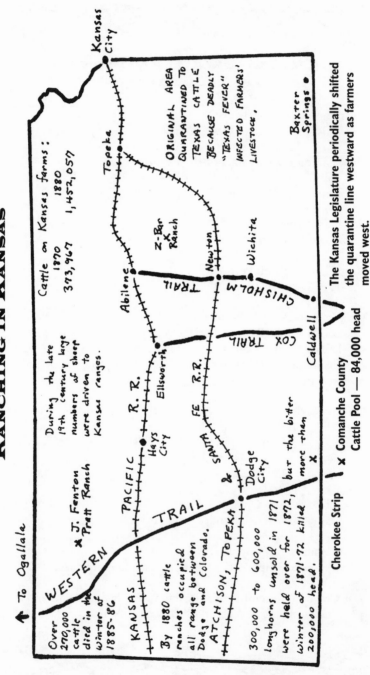

RANCHING IN KANSAS

To Ogallala

WESTERN TRAIL

KANSAS PACIFIC R. R.

SANTA FE R. R.

ATCHISON, TOPEKA &

★ J. Fenton Pratt Ranch

Hays City

Dodge City

Ellsworth

Abilene

Topeka

Kansas City

Newton

Wichita

Caldwell

CHISHOLM TRAIL

COX TRAIL

z-Bar x Ranch

Baxter Springs ●

Cattle on Kansas farms:

1870   1880
373,967   1,452,057

During the late 19th century large numbers of sheep were driven to Kansas ranges.

ORIGINAL AREA QUARANTINED TO TEXAS CATTLE BECAUSE DEADLY "TEXAS FEVER" INFECTED FARMERS' LIVESTOCK.

Over 270,000 cattle died in the winter of 1885-86

By 1880 cattle ranches occupied all range between Dodge and Colorado.

300,000 to 600,000 longhorns unsold in 1871 were held over for 1872, but the bitter winter of 1871-72 killed more than 200,000 head.

Cherokee Strip

x Comanche County Cattle Pool — 84,000 head

The Kansas Legislature periodically shifted the quarantine line westward as farmers moved west.

Kansas

HILL SPRING RANCH

Stephen F. Jones, who founded the Hill Spring Ranch and erected a magnificent headquarters complex, arrived in Kansas in 1876 "with money sticking out of every pocket."

Jones was born in 1826 in Nashville, Tennessee, and in 1849 he and his bride, Louise, migrated to Texas. He became a stockraiser, moving to Colorado in 1868. Success in the booming western cattle business brought him in 1876 to the Flint Hills country of Kansas. Jones bought up several farms north of Strong City in Chase County, then shipped in 2,000 head of cattle by rail to stock his new "Deer Creek Ranch." Because water was supplied by a spring on the hill where his ranch was centered, Jones soon changed the name to "Hill Spring Ranch." Jones also invested in the lumber business and in the Strong City Bank, which he served as president.

In 1881 Jones built a ranch house suitable to his station in life. The three-story, eleven-room limestone structure was erected on the side of a hill two miles north of Strong City. There were stone terraces and stairways leading uphill to the mansion, as well as wrought-iron fences and a terraced rose garden with a spring-fed fountain. The limestone outbuildings included a storm cellar, con-

133

In 1881 Stephen F. Jones built a three-story ranch house with a mansard roof. The striking stone structure still stands, facing east on a hill two miles north of Strong City.

A three-story stone barn was built south of the house. A wagon and team could be driven up the ramp (at right) to the third-story hayloft.

nected by tunnel with the kitchen because Louise Jones feared Kansas tornadoes.

South of the house Jones raised a three-story limestone barn. The structure measured 110 feet by 60 feet, and two wide ramps led from ground level to the top-floor hayloft. There was an icehouse and a spring house, and even the chicken house was built of stone.

In 1888 Jones sold his 7,500-acre ranch to the Lantry brothers of Strong City. Jones moved his family to Kansas City, where he erected a fine house of white limestone quarried in Chase County. The Lantrys did not move from town into the towering ranch house, but they added 5,500 acres to the spread.

Charles H. Patton of Reading, Kansas, purchased the 13,000-acre ranch in 1907. Ten years later he sold the stone buildings and 1,100 acres to Otto Benninghoven, and in 1921 Lester Urschell bought the balance of the range. In 1935 George Davis purchased both tracts and began operating the old ranch as a whole.

After twenty years, Davis found it necessary to take on two partners, who owned a grain company in Kansas City and rangeland

CATTLE POOLS

Although Stephen F. Jones was wealthy enough to build a Kansas ranch by purchasing several farms, most cattlemen found it too expensive to put together a large spread from land coveted by farmers. During the early years of the drives from Texas, herds that arrived at Kansas railheads when prices were unfavorable often were held over on the open range until the next season. But uncharacteristic amounts of rainfall from 1875 to 1885, along with various irrigation projects, encouraged farmers to move even to western Kansas. Ranchers in the southern part of the state leased range across the border in the Cherokee Strip, but that vast area was opened to homesteaders in 1893.

For several years, however, Kansas cattlemen managed to operate on a large scale by pooling their herds and resources. At Ellis in February 1878 twenty-six ranchers owning a total of 11,000 cattle divided their region into range districts and organized an annual spring roundup. In 1882 the Smoky Hill Pool was formed to control grazing along a thirty-by-twelve-mile range adjoining the Smoky Hill River. Within two years there were 15,000 head of cattle, and the spring roundup required 150 riders and 600 horses. During this period the Pawnee Valley Stock Breeders' Association boasted a capital stock of $200,000; headquartered in the buildings of old Fort Larned, this organization specialized in registered Shorthorn bulls, Clydesdale stallions — and sheep.

The largest of these groups was the Comanche Cattle Pool, organized in 1880 across most of Comanche County and in parts of Barber County and the Cherokee Strip. An office was maintained in Medicine Lodge, and at one point more than a score of cattlemen grazed an estimated 84,000 head inside Comanche fences. The Pool spent $30,000 on an enclosure that stretched a total of 180 miles and was built of 60,000 fence posts and 240,000 pounds of barbed wire. But after the disastrous winter of 1885-86, only 7,000 cattle were still alive inside the fence, and the Comanche Cattle Pool — along with most of the other Kansas pools — was dissolved.

Cowboys washing off trail dust in a Kansas creek.
— Courtesy Kansas State Historical Society

in Comanche and Barber counties. They used a Z Bar brand in Comanche County, and Davis and his partners named their new corporation the Z Bar Cattle Company.

In 1971 the Z Bar's 1881 ranch house was added to the National Register of Historic Places. The National Park Trust bought the Hill Spring/Z Bar Ranch in 1994, leasing the grazing land to pay property taxes and to accumulate funds to develop the old Hill Spring Ranch into a Flint Hills historic site and game preserve.

COTTONWOOD RANCH

It is common knowledge that Englishmen were attracted in large numbers to western ranching. One pocket of English settlers clustered in Sheridan County in northwestern Kansas, where the ranch buildings of John Fenton Pratt still stand as a tangible reminder of a significant chapter of frontier ranching.

Abraham Pratt of Yorkshire County, England, immigrated to the United States in 1878, first trying Nebraska, then filing a homestead claim in Sheridan County. A widower with four grown children, Pratt had ventured to California as a young man, then three decades later sold his businesses in England to return to the West. By 1880 Pratt's eldest son, John Fenton, had joined him in his dugout. "Fent" Pratt homesteaded a claim on the north side of the South Solomon River, adjacent to his father's property. Within the next few years Fent's younger brother and two family friends also settled in Sheridan County. In 1885 Fent built a one-room stone house, where the five bachelors suffered through the bitter winter of 1885-86.

The "Big Die-Up" during that winter devastated cattle ranchers, who already had severely damaged the open range by overgrazing their herds. With the elimination of a great many cattle ranchers, sheepraisers were able to expand their open-range operations. Sheep could graze ranges no longer fit for cattle, and sheep provided two annual payoffs — the lamb crop and the wool fleece — rather than one, as cattle did. The Pratts and other English settlers of Sheridan County came from a sheep-raising culture, and they readily turned to the cultivation of woollies. Although most of western Kansas had been cattle country since the post-Civil War era, for the next decade and a half Sheridan County and other parts of the region raised large flocks of sheep.

The Pratts ran a combined flock of about 1,600 sheep, the majority owned by Fent. Fent also was active as an investor and financier, while his brother ran his own farming operation, and Abraham owned the lumberyard in nearby Skelton (present-day Studley). The family had the capital to pursue various business enterprises and to expand their ranch headquarters.

To the one-room stone house had been added sod outbuildings and corrals. But during the early 1890s English workmen were employed to expand the house and to replace the sod structures with handsome stone construction. The result was a striking headquarters complex strongly reminiscent of Yorkshire County.

The enlarged house featured a front porch and two bay windows. A low stone wall surrounded the residence and stone storage building. Behind the Pratt home, 120 feet to the north, stood a line of outbuildings connected by stone walls. The stable was flanked by

Rambling front porch of the Pratt ranch house.

Stone barn, standing 120 feet behind the ranch house.

A barn (left) and stable are connected by stone walls.

An auxiliary building behind the Pratt ranch house.

a barn and a large shop, and the walls of these buildings and the connecting walls formed a major part of the corrals. Building stone was
quarried from Pratt land one and a half miles away. (Other English
settlers in the area also utilized the Pratt quarry.)

After Abraham Pratt died in 1901, the Pratt brothers divided
their father's estate with their sisters in England, then pursued their
own interests. Fent sold his sheep and most of the land of
Cottonwood Ranch, except for his original homestead. He used the
money from these sales for investment capital.

Fent was married on the last day of 1888 to his sweetheart
from England, Jennie Place, who journeyed to Kansas to be wed.
The couple had two daughters. Fent died at eighty-one in 1937, and
Jennie lived until her ninety-eighth year, in 1959. Their younger
daughter, Elsie (1894-1975) was married, but Hilda (1889-1980)
never wed and lived at Cottonwood Ranch until 1978. The State of
Kansas purchased twenty-three acres of Cottonwood, including the
stone buildings, in 1983, and the headquarters complex is gradually
being developed into a living museum.

KANSAS CATTLE FRONTIER IN A CORRAL

Biggest Cattle Operation: Comanche Cattle Pool
(84,000 head under 20 brands)

First Rodeo: Caldwell "Grand Cowboy Tournament" —
May 1, 1885

Best Cattle Towns: Dodge City, 1876-85
Abilene, 1867-71
Wichita, 1872-76
Caldwell, 1880-85
Ellsworth, 1871-73
Newton, 1871-72

Most Popular Cattlemen's Hotels:
Drover's Cottage — built in Abilene in 1868 by J.G.
McCoy, then moved to Ellsworth in 1872; 75 rooms
Dodge House — Dodge City; 38 rooms
Southwestern Hotel — Caldwell; three-story brick, 38
rooms, ladies' parlor
Leland Hotel — Caldwell; three-story brick, 50 rooms,
view of the Cherokee Strip

Best Saloons: Alamo (Abilene)
Long Branch (Dodge City)
Red Light (Caldwell)
Reno House (Wichita)
Gold Room (Newton)
Bull's Head (Abilene)
First Chance/Last Chance (early Caldwell
on the Oklahoma/Kansas border)

Deadliest Saloon Fight: Perry Tuttle's Dance Hall in Newton
(August 20, 1871), Texas cowboys vs. railroad men. Four
dead and five wounded.

5

Nebraska

SCOUT'S REST

Buffalo Bill Cody was a teamster, trapper, Pony Express rider, buffalo hunter, army scout, hunting guide, Wild West show empresario — and rancher. Cody ran a pioneer open-range operation in Nebraska's Sand Hills; he owned Scout's Rest, a showplace ranch just outside of North Platte; and later he put together a cattle ranch in northern Wyoming. Most significantly, Buffalo Bill's Wild West was instrumental in establishing the cowboy as a folk hero of mythic proportions.

By 1869 Cody, serving as chief of scouts of the Fifth Cavalry, was stationed at Fort McPherson near North Platte, Nebraska, where he moved his family into a log cabin. A few years later he launched a career as a stage performer, and in 1878 Cody sent money to his wife, Louisa (usually called "Lulu"), for the purchase of land at North Platte. Louisa paid $750 for 160 acres adjacent to the railroad tracks on the outskirts of North Platte. There was a small house, along with a corral and outbuildings and lush pastureland. Just south of the tracks Lulu supervised construction of a story-and-a-half frame residence which became known locally as the "Welcome Wigwam."

Cody returned to the new house from a profitable stage tour,

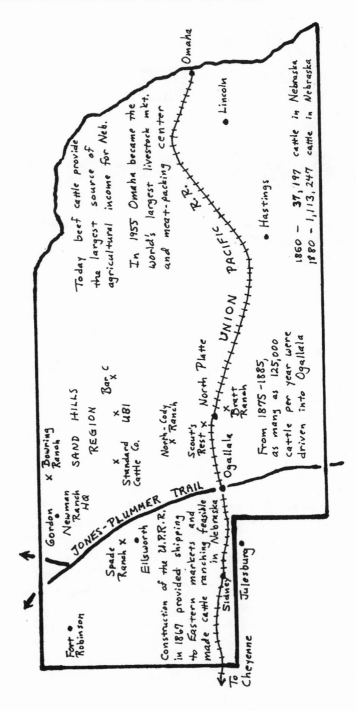

RANCHING IN NEBRASKA

Fort Robinson

Gordon • × Bowring Ranch
Newman Ranch HQ
SAND HILLS REGION

Spade Ranch ×
Ellsworth

Standard Cattle Co. × × UBI

Bar C ×

North-Cody × Ranch

Scout's Rest ×
North Platte ×
Bratt Ranch ×

Ogallala

JONES-PLUMMER TRAIL

Sidney •
Julesburg •
To Cheyenne

Omaha •
Lincoln •
Hastings •

UNION PACIFIC R.R.

Today beef cattle provide the largest source of agricultural income for Neb.

In 1955 Omaha became the world's largest livestock mkt. and meat-packing center

Construction of the U.P.R.R. in 1869 provided shipping to Eastern markets and made cattle ranching feasible in Nebraska

From 1875-1885, as many as 125,000 cattle per year were driven into Ogallala

1860 - 37,197 cattle in Nebraska
1880 - 1,113,247 cattle in Nebraska

Buffalo Bill Cody, North Platte's most famous citizen.

but he would spend most of his time at his Sand Hills ranch, which had been located by his partner, Frank North. A large-scale roundup was conducted there, and Cody was so deeply impressed by the riding and roping skills of the cowboys that he soon would feature these activities in his show. The convivial Cody quickly recognized the drudgery and toil of the cowboy's routine. "As there is nothing but hard work on these roundups, having to be in the saddle all day, and standing guard over the cattle at night, rain or shine, I could not possibly find out where the fun came in that North had promised me."

The partners maintained their Sand Hills ranching venture only until 1882, but Cody continued to send money for land purchases to North Platte. The property took shape west of town and north of the railroad tracks, eventually totaling 4,000 acres. Keenly aware of her husband's generous and impulsive nature, Lulu discreetly put the property in her own name, which later would greatly distress Cody, particularly since the couple was often embattled. Lulu preferred to stay at her Welcome Wigwam in town, where she was active socially, and Cody continued to develop the ranch he optimistically dubbed "Scout's Rest."

Cody moved his sister Julia, her husband, Al Goodman, and their seven children to Scout's Rest in 1885. Al was employed as foreman of Scout's Rest, and during the years that he managed the ranch there were steady profits. Cody assigned Al and Julia a major task in 1886: construction of a two-story, nine-room house with a towering cupola. Al engaged a contractor who agreed to build the house within six months for $3,500. Julia was in charge of designing and furnishing the big residence. Cody stipulated that the porches should be ten feet wide, instead of the customary six, and that his upstairs bedroom should include "a side board [with] some

The log cabin built in 1877 as headquarters of the Cody-North ranch now is on display at Scout's Rest.

CODY'S DISMAL RIVER RANCH

From 1877 to 1882 Buffalo Bill Cody and fellow scout Frank North were partners in an open-range cattle ranch along the Dismal River about sixty-five miles northwest of North Platte. While Cody was on tour with his Wild West show in 1877, North built a log ranch house, sod stable, and a cedar pole corral. The isolated headquarters site chosen by North was at the headwaters of the South Fork of the Dismal, deeper into Nebraska's Sand Hills than any cattleman had ever ventured.

In Omaha in July 1877 Cody and North purchased 1,500 head of Texas longhorns, hired a crew, and drove their herd to the Dismal. Cody was especially impressed by one of the Texas drovers, tall Buck Taylor, who later would star in Buffalo Bill's Wild West as the "King of the Cow-boys."

Although the partners ran cattle along their Dismal range for the next few years, they never filed a homestead claim or purchased any land. In 1882 Cody and North sold their cattle, buildings, and range rights to rancher John Bratt for $75,000.

THE "OLD GLORY BLOW OUT"

While visiting his family in North Platte during the summer of 1882, Buffalo Bill Cody organized a landmark Fourth of July celebration. The event coincided with the last large-scale open range roundup in western Nebraska.

Thousands of handbills were printed and distributed widely, proclaiming cash and merchandise prizes for winners of riding and roping events. At 10:30 A.M. an "immense crowd" watched a band lead an opening parade through the streets of North Platte to the racetrack. Cody had gathered several buffalo and a small herd of Texas longhorns for roping exhibitions, and after being lassoed these wild beasts were ridden, to the delight of the spectators.

Everyone had a fine time, and Cody took note of the enthusiastic crowd reactions. His creative imagination stimulated by the "Old Glory Blow Out," Cody immediately began organizing a company that would take a Wild West show on the road. The 1882 event was an important step in the popularization of rodeos and the development of the most famous and successful of the traveling Wild West shows.

nice decanters & glasses." This impressive home was a tourist attraction even while under construction, and after it was completed Cody genially hosted a constant parade of guests. When he was at Scout's Rest a flag was hoisted, and visitors often found him on his large porch, happy to offer drinks and other hospitality. A gracious host, even during his long absences, Cody sent written instructions for Goodman: "Tell Al to always keep a little Busine [whiskey] in the house — and keep open house to any nice callers."

Soon an enormous barn was erected: 148 feet long, 70 feet wide, and 40 feet tall. Four-foot letters on the roof spelled out "SCOUT'S REST RANCH" and could be seen from passing railroad cars more than a mile to the south. In 1891 a large, T-shaped barn was constructed (this structure burned in 1904). There was also a pond with a bridge labeled "SCOUT'S REST RANCH" in raised letters,

This side view of the Scout's Rest Ranch house, built in 1886, shows the rear addition, added in 1909.

Cody's upstairs bedroom.

Cody's 148-foot-long barn sported a sign that could be read by railroad passengers more than a mile away.

Outbuildings clustered just west of the main house include the spring house in the foreground, the cook house to the left, and the icehouse in the rear.

Stalls at the Scout's Rest barn.

Buffalo Bill's saddle, ca. 1877; his name is carved behind the cantle.

The enormous loft at the Scout's Rest barn.

and numerous outbuildings, including an icehouse and a spring house. An *Omaha Herald* reporter described Scout's Rest in detail, declaring the ranch "one of the finest on the continent."

Cody built a twelve-mile-long irrigation ditch, and he watered more than 1,000 acres, growing corn, alfalfa, and oats. There were eighty work horses and thirty employees, a number that doubled during certain seasons. Cody (or actually Al Goodman) raised thoroughbred horses and blooded cattle (the ranch boasted a prize Hereford bull named Earl Horace). Between touring seasons Cody sometimes refitted the vehicles and worked and trained the animals of his Wild West show. At all times of the year members of his troupe could be found at Scout's Rest, usually recuperating from injuries. When Annie Oakley was on the ranch there were shooting contests, and a racetrack on the property was frequently in use.

Cody loved to drive into town in a tallyho, which he had purchased while on tour in England and shipped back to Scout's Rest. Cody would get the mail, visit a saloon and meet the train, bringing guests to the ranch in his grand vehicle. Visitors included show people, prominent politicians, army officers, European royalty, and old friends from his frontier days. Scout's Rest was always lively when Cody was there, and he often brought excitement to North Platte, going on legendary sprees in the town's saloons.

The Welcome Wigwam went up in flames in 1891, and Mrs. Cody soon moved into a more ornate Welcome Wigwam in North Platte. In 1892 Al and Julia Goodman moved their family back to their former home in Kansas, and future management of Scout's Rest would never again be as effective. Severe financial losses suffered during the Panic of 1893 caused Cody to place a heavy mortgage on Scout's Rest. By 1895 increasing alienation between Cody and Lulu caused him to begin moving cattle from Scout's Rest to Wyoming's magnificent Big Horn area, where he established the TE Ranch.

As Cody began organizing another ranch, he acquired a herd with a TE brand, which provided a name for the new spread. In time the TE expanded to 4,600 acres, and Cody controlled other nearby properties. He also pursued other interests in the region, including the founding of a town named Cody. (Today the TE still is in operation, headquartered in original structures, and the Irma Hotel in Cody and a picturesque hunting lodge east of Yellowstone Park are among other Cody landmarks in the region.)

JOHN BRATT'S
DOUBLE O RANCH

Shortly after coming to North Platte as an army scout, Cody established a lifelong friendship with pioneer Nebraska cattleman John Bratt. Born in England in 1843, Bratt migrated to the United States in 1864 and became a teamster in the West. After making several freighting trips from Omaha to Fort Laramie, he came to Fort McPherson in 1869 to fill a government hay contract. By the next year Bratt had entered the cattle business, establishing his headquarters four miles southeast of North Platte, near the fort.

Bratt built a sod house with loopholes for riflemen, along with a set of sod barns with a cedar pole corral, and he opened a meat market in North Platte. The Bratt and Company brand would be a Double O. Bratt established a vast range from the Platte River south to the Republican, and from Fort McPherson on the east to Fort Sedgwick, in northeastern Colorado. "We ranged ten to fifteen thousand cattle at times and about one thousand horses and mares," reminisced Bratt, "branding three thousand to five thousand calves and two hundred to three hundred colts per year."

A sweeping prairie fire burned all of the range south of the Platte in 1874, forcing Bratt to move his herd north into the Sand Hills, open country still roamed by Indians. But grazing was excellent, and Bratt expanded into the area. In 1882 he paid $75,000 to Cody and Frank North for their Dismal River Ranch. Three years later he purchased from the Union Pacific 123,673 acres around Birdwood Creek, and fenced in this large range about one hundred miles above North Platte. Active in public affairs, the noted old rancher died in 1918, a year after Cody.

In 1909 Lulu Cody moved into the big Scout's Rest ranch house, along with assorted relatives and friends. A major addition to the rear expanded the house to nineteen rooms, and a busy social schedule continued at Scout's Rest, despite the absence of Buffalo

Bill. Scout's Rest was the site of a reconciliation between Bill and Lulu in 1910, on the occasion of the wedding of a granddaughter.

But by the next year pressing debts caused Cody to sell Scout's Rest to Pawnee Bill (Gordon Lillie), who had merged his show with Buffalo Bill's Wild West in 1908. The property title still was in the name of Louisa Cody, who agreed to a price of $100,000. Lulu received $80,000 of the sale price, while $20,000 went toward Cody's debts to Pawnee Bill. Lulu continued to live in the old house for a time, and the sale was not made public until 1913. Cody continued to tour until he died at the age of seventy in 1917.

Ten years later, Scout's Rest was purchased by Henry Kuhlman, who moved his family into the house and successfully raised Polled Herefords. The old house eventually deteriorated to the point that the Kuhlmans built a new home to the west, leaving the old mansion empty. But tourists had never stopped asking to see Buffalo Bill's ranch, and in 1960 North Platte and the county purchased the decaying house and barn, along with twenty-five acres, for $75,000. The State of Nebraska agreed to put up half of the money, because restoration costs would far exceed the purchase price. Scout's Rest opened as a State Historical Park in 1965, and Buffalo Bill's charming headquarters site still is Nebraska's most popular ranch.

 ## THE SPADE RANCH

Bartlett Richards, a native of Vermont, sought his fortune in western cattle ranching, moving to the Wyoming frontier as a determined eighteen-year-old in 1877. By 1884 he controlled or managed herds on several ranges under a dozen brands, and he started a bank at Chadron, in northwestern Nebraska. Bartlett served as president and his older brother, DeForest (later to become governor of Wyoming), was vice-president. Undeterred by government orders to pull down fencing around 61,000 acres of public land, Bartlett was equally undiscouraged by the devastating winter of 1885-86.

Recognizing the need for winter hay, by 1888 he had decided to build a cattle empire in Nebraska's nearly uninhabited but grassy Sand Hills. Bartlett originated the Spade Ranch in Sheridan County,

Chow time on the range near Ellsworth, early 1900s.

Dipping cattle at the Ellsworth pens in the early 1900s.

A newspaper reporter at Spade Ranch chuckwagon in 1918.

adapting an Ace of Spades brand that would be difficult for rustlers to alter. He hauled in a few structures from the abandoned Newman Ranch on the Niobrara River, as well as a small cabin from one of his own operations to the west. Even his headquarters was on public domain, and once again he began fencing land that he did not own.

Soon Bartlett controlled a range of approximately 500,000 acres. He imported Hereford bulls, and his herds grew to more than 40,000 head of improved cattle. Bartlett acquired large beef contracts for Indian reservations, with permission to graze up to 10,000 contracted cattle on reservation lands.

Bartlett engaged his brother, Jarvis, as general manager and purchasing agent, while DeForest handled financing from the Bank of Chadron. Tongues wagged when Bartlett married his own niece, DeForest's pretty daughter, but the couple happily reared a family.

The Burlington Railroad laid tracks and built a depot and water tower twenty miles south of Spade Ranch headquarters, and in about 1890 Bartlett constructed a store, shipping pens, business office, and hotel. In 1902 he erected a residence in Ellsworth. Telephone lines were built from Ellsworth and Chadron to headquarters, and water wells were drilled across Spade range.

In 1890 Bartlett Richards built a store and ranch office in Ellsworth on the Burlington Railroad.

SAND HILLS RANGE

The Sand Hills cover 19,300 square miles of Nebraska and lower South Dakota, forming the largest area of sand dunes in the western hemisphere. The Dismal, Snake, Calamus, and Niobrara rivers wind gently through pine-studded canyons, while nearly 1,000 small lakes dot the vast expanse of grass-covered sand hills. The egg-shaped region stretches 250 miles north to south and as much as 100 miles east to west.

Although the grassy, well-watered region teemed with deer, antelope, elk, and other prairie wildlife, settlers moving up the nearby Oregon Trail ignored this forbidding portion of the "Great American Desert." Early ranchers also avoided the Sand Hills, assigning line riders to keep cattle out of a rugged area that promised to be difficult at roundup time. But longhorns being trailed north, as well as cattle from nearby herds, inevitably strayed into the Sand Hills, then thrived and multiplied. When a Newman Ranch cowboy rode in search of stray horses, he found large numbers of sleek mavericks. A roundup crew was sent into the Sand Hills and came out with a herd that filled several cattle cars bound for the Chicago stockyards.

One of the finest cattle ranges in the United States had been discovered, and cattlemen soon put together ranches as large as 500,000 acres, although most of their land was public domain. The Spade Ranch built a herd of more than 40,000 Herefords; to the east of the Spade the Standard Cattle Company ran 18,000 head; and still further east were other big ranches, the British-owned UBI and the Bar C. Homesteaders moved in, but the Sand Hills proved inhospitable for farmers. For more than half a century the Sand Hills area has been cattle country, and there still are ranches as large as 100,000 acres.

The Spanish-American War of 1898 created a great demand for beef, causing Bartlett to expand his operations by organizing the Nebraska Land and Feeding Company in partnership with Englishman William G. Comstock. But veterans who went West to settle on their "land bonuses" found great expanses of public domain fenced off by cattlemen like Bartlett Richards. When famed Rough Rider Theodore Roosevelt became president, he launched federal investigations, and late in 1905 Bartlett Richards and Will Comstock were found guilty of illegally fencing 212,000 acres of government land. The partners were fined $300 cash and sentenced to six hours in the custody of their Omaha attorneys, time that was spent enjoying a champagne dinner. Bartlett and his family then departed for a vacation in Europe.

Roosevelt was incensed when he read newspaper criticisms of the outcome of the Richards-Comstock trial, as well as similar light punishments of other ranchers. Agents of the land office, under Secret Service protection, investigated the actual extent of Spade ranges. Another land-fraud trial followed, featuring hundreds of government witnesses, and Richards and Comstock again were convicted, along with other employees of the Nebraska Land and Feeding Company. While these convictions were under appeal, the winter of 1909-10 badly hurt the Spade herd. Then Spade foreman Marquard Petersen opportunistically filed a homestead claim on the headquarters complex area, moving into the Spade buildings despite being fired by the owners.

Appeals went all the way to the U.S. Supreme Court, but the convictions were upheld. Although Richards, Comstock, and several associates were sentenced to a year's imprisonment, the millionaire inmates were permitted a choice of institutions. The county jail at Hastings was selected, where they were made comfortable with rugs and curtains, books, magazines and newspapers, and quality meals prepared by a Japanese cook. Mrs. Comstock moved to Hastings to be near her husband, but Bartlett Richards adamantly refused to allow his wife, children, or mother to see him in jail. His health failed while he was behind bars, and he died at the age of fifty-two on September 5, 1911 — just one month before his scheduled release. Will Comstock died only a few years after his release from the Hastings jail.

The Spade Ranch had been reduced to merely 22,000 acres.

The Bixby family, associates of Bartlett Richards since 1908, acquired the Spade for the price of the mortgage. The buildings at the Spade Ranch headquarters, homesteaded by Marquard Petersen, eventually were placed on the National Register of Historic Places. The store built by Bartlett Richards in Ellsworth more than a century ago is still in operation.

99   BOWRING BAR 99 RANCH

During the 1870s John Bratt and other pioneer cattlemen began to penetrate the isolated but rich grazing lands of the Sand Hills, and during the 1880s big ranchers overstocked this enormous open-range area. But disastrous winters during the mid-1880s drove many large operators out of business. Smaller ranches began to dot the Sand Hills, usually begun by men who started with a 160-acre homestead.

Henry Bowring, a native of England, moved his large family in 1885 to a Sand Hills homestead near Gordon. When the seventh Bowring son, Arthur, turned twenty-one in 1894, he filed his own homestead claim just north of Merriman. Arthur built a sod house with a wooden floor, and began ranching with Shorthorn cattle. Later he switched to Herefords, using a Bar 99 brand.

Bowring began purchasing additional tracts, usually from farmers or small ranchers ready to quit the Sand Hills and sell out. He picked up an additional 480 acres of government land through the Kinkaid Act of 1904, and eventually his ranch encompassed 7,202 acres.

In 1908 the thirty-five-year-old bachelor married a Merriman schoolmarm, Anna Mabel Holbrook; however, the next year Anna and her infant died in childbirth. For the next two decades Bowring continued to live alone, building up his ranch and engaging actively in public affairs. He served as a justice of the peace, road overseer, election judge, county commissioner, school board member, deputy of the Nebraska Game and Fish Commission, and state legislator.

When in his fifties Arthur met Eve Kelly Forester by offering aid when her automobile broke down. She was a traveling sales representative for a Lincoln company, a daring occupation for a woman

The sprawling Bar 99 ranch house today displays the glassware and silver collected by Eve Bowring during her travels.

Barn at the Bar 99.

Bar 99 bunkhouse.

during the 1920s. Eve, an attractive and energetic widow with three sons, married Arthur in 1928. The family lived in a frame house near the 1894 soddy, barns, bunkhouses, corrals and various outbuildings.

In 1944, after working the Bar 99 for half a century, Arthur died at the age of seventy-one while helping a neighbor fight a fire. By this time Eve had proved herself as active in public life as her husband. After becoming a leader in the local Republican Party, she was a member of the National Health Institute and traveled widely as a longtime member of the U.S. Parole Board. In 1954 she was asked to fill the remainder of the term of a deceased Nebraska senator, serving seven months in the U.S. Senate. Despite extended absences, Eve always was at the ranch to work during calving time, and otherwise helped to maintain the Bar 99.

Eve Bowring survived each of her sons, as well as two husbands, and late in life she determined to preserve the Bar 99 as a living history museum of a working ranch. She specified that the Hereford bloodline of the Bar 99 herd be maintained. Eve died in 1985, one day after her ninety-third birthday. The ranch buildings were preserved, a replica of the original soddy and a visitor center were constructed, and today the Arthur Bowring Sandhills Ranch State Historical Park documents the developments of cattle ranching in Nebraska over the last century.

Replica of the 1894 sod house built by Arthur Bowring.

NEBRASKA
RANCHING IN A CORRAL

Biggest Ranch:
Spade (500,000 acres; 40,000 cattle)

First Rodeo:
Old Glory Blow Out, North Platte — July 4, 1882

Best Cattle Town:
Ogallala — "Cowboy Capital of Nebraska," 1875-85

Most Popular Saloons:
Cowboy's Rest, Ogallala
Crystal Palace, Ogallala

Most Famous Ranch Animal:
Earl Horace, pedigreed Hereford bull, Scout's Rest

Union Stock Yards:
Organized in 1883 in Omaha by cattle baron Alex Swan and other investors, eventually growing into world's largest.

Kinkaid Act (1904):
Moses Kinkaid (1856-1922) was a Nebraska senator (1902-22) who authored a statute increasing the size of free homesteads from 160 acres to 640 acres in 37 semiarid counties of western Nebraska. Farmers who settled on Kinkaid claims soon found farming in this area unprofitable, and ranchers eventually acquired most of these tracts.

6

Colorado

JOHN WESLEY ILIFF
Cattle King of the Plains

Colorado's first great ranching empire was built by "the squarest men that ever rode over these Plains," according to Wyoming cattle baron Alexander Swan. Born in 1831 in McLuney, Ohio, John Wesley Iliff was named after the founder of Methodism, and for three years attended Ohio Wesleyan University. His father prospered as a stock farmer, taught John Wesley the cattle business, and offered him $7,500 to buy an Ohio farm. John Wesley instead asked for just $500 — as a stake to go West.

John Wesley Iliff journeyed in 1856 to Kansas, where he helped found the town of Princeton, then engaged in farming and mercantile enterprises. Three years later he was attracted to the Colorado Gold Rush along Cherry Creek, where Denver City was developing. With money from the sale of his Kansas interests, Iliff took an ox-train of provisions to Denver and opened a mercantile establishment. But soon he began to buy cattle and fatten them on nearby open ranges, then sell them to mining camps and Denver butcher shops.

Iliff aggressively pursued beef contractors with army posts, Indian reservations, and Union Pacific construction crews. To fill these contracts he rapidly built a herd and ranged it in northeastern

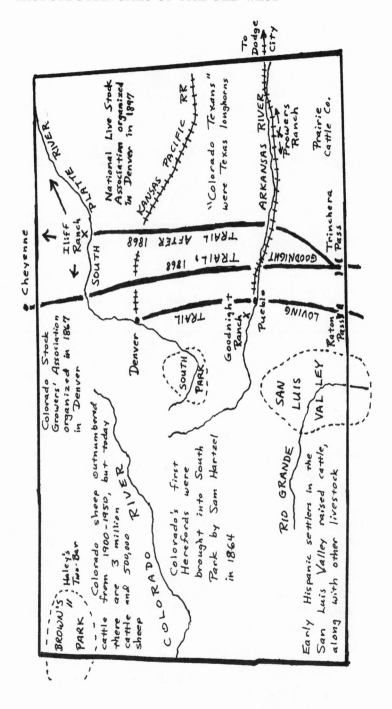

RANCHING IN COLORADO

Colorado along the South Platte River. Iliff purchased 800 of the first herd driven into Colorado by Oliver Loving in 1866, and later bought 10,000 to 15,000 Texas longhorns per year. Buying these animals at $10 to $15 per head, he fattened them for a year or two, then sold for $30 to $37 each. In 1874, for example, Iliff's costs for wintering and herding 26,000 cattle totaled just $15,000, while sales added up to $128,000. During that year Iliff employed, during different seasons, 12 to 35 men, and owned 120 horses.

Such profits allowed him to absorb losses such as those suffered during the severe winter of 1871-72. A canny judge of cattle, Iliff produced large profit margins by shrewd trades and purchases, as well as by energetic attention to the details of his business. Twice married, he was a devoted family man, but his first preoccupation was with his cattle empire.

During peak years Iliff's herds numbered at least 35,000 head, and he utilized a range extending more than 100 miles westward along the South Platte from Colorado's eastern border, and sixty miles northward into Wyoming. He bought only enough of this land to secure control of the surrounding grasslands. At his death, in addition to Denver real estate properties, he owned title to 15,558 acres of rangeland in 105 parcels. There were fifty 160-acre tracts; twenty-seven tracts of eighty acres; forty-four tracts of just forty acres; and others as small as fourteen acres. These tracts were strategically placed for ranch headquarters and line camps, and, of course, to control water sources. In 1877, when he owned more than 21,000 cattle, his property taxes were merely $3,062.60.

During the 1870s, Iliff upgraded his cattle by bringing in Shorthorn bulls from the East. At different times he grazed his cattle on free grass in Wyoming, Kansas, Texas, Oklahoma and Nebraska, as well as in Colorado. But the unrelenting press of business activities, as well as the chain of cigars he incessantly smoked, undermined his health. The "cattle king of the Plains" died in 1878 at the age of forty-six, after amassing a fortune in less than two decades of open-range ranching.

JOHN WESLEY PROWERS
Colorado's Other Cattle King

Like John Wesley Iliff, John Wesley Prowers was named after the founder of Methodism; like Iliff, Prowers lived only to his forty-sixth year; and also like Iliff, Prowers built an enormous cattle empire in Colorado.

Born in Missouri in 1839, Prowers headed west at the age of eighteen and found employment as a clerk at Bent's Fort on the Arkansas River in southern Colorado. During seven years of service to famed Indian trader William Bent, Prowers made more than a score of trips across the Plains, usually leading supply trains. In 1861 he married the daughter of a Cheyenne chief, the same year that he purchased 600 head of Missouri cattle and drove them to Colorado. He bought land near Fort Lyon in 1863, herding government horses, cattle and mules, in addition to his own cattle.

By 1865 Prowers had entered the freighting business, and three years later he began farming at Boggsville, which soon became the county seat when Bent County was organized. Prowers was appointed a county commissioner, beginning a secondary career in public service that led to the Colorado Legislature. He moved to Las Animas in 1873, engaging in various businesses, becoming a bank president, and establishing a large slaughterhouse.

Prowers became noted for his vast ranching activities, controlling forty miles of river frontage along the Arkansas and ranging his cattle on 400,000 acres. He crossed his Missouri cows first with purchased Shorthorn bulls, then with Herefords. In 1871 he bought a cow named Gentle the Twelfth, and during the next ten years this fine animal produced offspring which netted him $10,800. With a herd consistently numbering 10,000 or more, Prowers made large shipments of beef to the East.

At the height of his fame as a cattleman and politician, Prowers fell ill late in 1883. He traveled to Kansas in search of medical aid, but died on February 14, 1884.

PAT CHARLES GOODNIGHT'S ROCK CANYON RANCH

Legendary pioneer cattleman Charles Goodnight blazed the Goodnight-Loving Trail across West Texas into New Mexico in 1866. His partner, Oliver Loving, continued north with 800 head of longhorns through Raton Pass into Colorado, where he sold out to John Iliff. Although Loving was fatally wounded by Comanches in 1867, Goodnight shoved another, larger herd north and established the first big cattle ranch in southern Colorado.

Goodnight built a log headquarters at the head of Apishapa Canyon, forty miles northeast of Trinidad in the midst of isolated grazing country. In 1868 Iliff bought Goodnight's Apishapa cattle for delivery to Cheyenne, and the famous trail was extended into Wyoming. Iliff rapidly expanded his operations, purchasing nearly 30,000 head of cattle from Goodnight during the next three years. Goodnight drove herd after herd of longhorns into Colorado, shifting his trail to the east. He entered Colorado at Trinchera Pass rather than at Raton, where the old mountain man Dick Wootton charged a toll. Refusing to pay toll fees for as many cattle as he intended to drive, Goodnight blazed a new route into Colorado. Wootton finally offered Goodnight free passage, but cattlemen already were following the new trail.

By 1869 Goodnight had decided to lease the open range around Apishapa Canyon in favor of a new ranch location along the Arkansas River west of Pueblo. He preempted a grassy, well-watered mountain range in the shape of a triangle, with twenty-five miles of the Arkansas as the northern line. The St. Charles River flowing into the Arkansas defined the east, while Hardscrabble Creek was on the west. Goodnight established headquarters five miles west of Pueblo in a sheltered valley where the Arkansas cut through a narrow rock canyon. He registered a "PAT" brand for his Rock Canyon Ranch.

Goodnight married Molly Dyer in 1870, brought his bride to Rock Canyon Ranch, and placed his brother-in-law, Leigh Dyer, in charge of the outfit. Goodnight himself continued leading trail drives, after wintering thousands of cattle annually on "the Goodnight range." Substantial rock corrals were erected, and line camps were built on the St. Charles and at Babcock's Hole, on the

THE TELEGRAPH POLE IS FINE

In 1870 the thirty-four-year-old Goodnight traveled to Kentucky to marry his thirty-year-old sweetheart, Molly Dyer. After a hard return journey, the couple arrived at the Drovers' Hotel in Pueblo on the same night that vigilantes hanged two stock thieves from a nearby telegraph pole.

When the bride learned that two lynch victims were dangling near her hotel, she was visibly distressed. Goodnight, who had been plagued by these very rustlers, could offer little comfort.

"Well," he shrugged, "I don't think it hurt the telegraph pole."

Molly was doubly upset at her husband's callousness. Goodnight had to arrange a honeymoon excursion with friends to the Garden of the Gods and other spectacular sights of the Pike's Peak country, before taking her home to the Rock Canyon Ranch.

Hardscrabble. Goodnight ditched his valley for irrigation, began raising corn and other crops, and planted an apple orchard with trees brought in by stagecoach. He also invested heavily in Pueblo real estate and helped found a bank, before the Panic of 1873 "wiped me off the face of the earth."

Goodnight stuck with the Rock Canyon Ranch, and even joined John Prowers in establishing a meat-packing house in Las Animas. But the financial reverses of 1873 forced him to continue trail driving for the next couple of years, until a choice ranching opportunity beckoned in Texas. The Red River War had routed Comanches and Kiowas out of their refuge in Palo Duro Canyon in the Texas Panhandle. With these fierce horse warriors confined to reservations in Indian Territory, the sheltered, well-watered reaches of Palo Duro were open, and in 1876 Goodnight left Colorado to establish the JA Ranch, named after English investor John Adair. For a decade, however, Charles Goodnight had been instrumental in establishing the range cattle industry on the Colorado frontier.

ORA HALEY'S TWO BAR

During the early 1870s, a few cattlemen moved small herds into the sheltered grazing areas of Brown's Park in northwestern Colorado. But Brown's Park was a natural haunt for horse rustlers and cattle thieves, as well as an isolated hideout for fugitives from justice. And Utes, who inhabited a twelve-million-acre reservation in western Colorado, were renowned horse thieves and posed the danger of hostile raids as they roamed the entire area. In 1879 war-like Utes killed their agent, Nathan Meeker, and inflicted severe punishment on a cavalry column. An overwhelming army force re-established order, whereupon federal officials reclaimed the enormous Colorado tribal holdings and removed the Utes to two small reservations in Utah.

Ora Haley, who controlled a cattle empire in southern Wyoming, promptly moved across the line onto the ranges of northwestern Colorado. Early in the summer of 1880, Haley and A. W. Gregory, assigned to be foreman of the Colorado operation, led a trail drive which brought 7,000 head of cattle to the newest range of the Haley Livestock and Trading Company. Seeing the extent and quality of the grasslands, Haley then brought another 6,000 head in from Wyoming, thus establishing by far the largest herd under one ownership that had ever been assembled in north-western Colorado. The brand that he used was the Two Bar, applied to the left hip, with underbits in each ear.

Born in Maine in 1845, Ora Haley worked as a clerk in a general store until the outbreak of the Civil War, when he enlisted as a drummer boy. After the war he went west, finding employment as a bullwhacker with a train bound from Omaha to Denver. He spent the next couple of years bullwhacking with a freight outfit that operated between Denver and Central City. With his earnings he opened a butcher shop in Blackhawk, and when Union Pacific crews pushed westward across southern Wyoming in 1867, Haley opened another shop in Laramie.

After obtaining a contract to supply meat to Union Pacific crews, he bought his first ranch in 1871, about twenty miles south-west of Laramie. The bitter winter of 1871-72 killed many of his cattle, but the confident Haley always was philosophic about busi-

One of the abandoned log buildings at the
old Two Bar Ranch in Brown's Park.

The abandoned Two
Bar Ranch on the
Green River in
Brown's Park offers
a complex of empty
log structures.

Loading chute
on the old Two
Bar spread in
Brown's Park.

ness reverses. "I never look at my losses, or let them bother me," he would say. "I always look ahead."

Haley was elected to the Wyoming Territorial Legislature in 1871, and the next year he arranged a credit line with the First National Bank of Laramie. Also in 1872 Haley married Augusta Pfieffer of Omaha. The couple would become the parents of three daughters and a son, and in 1886 Haley built the finest home in Laramie for his family.

The rancher earned his first big cattle money in 1876, when he went to Oregon and put together a herd of meaty Herefords and Shorthorns for only $5 a head. He trailed these animals all the way to Omaha, where he hit a good market and sold out for $50 per head.

Haley believed that by increasing the number of cattle his operating expenses could be proportionally decreased, and he built up his ranching enterprise relentlessly. He began to acquire acreage in Wyoming that would give him control of surrounding range. Eventually, Haley owned 36,000 acres that gave him a string of ranches stretching twenty-five miles along the Laramie River, with the headquarters ranch at the confluence of the Big and Little Laramie rivers.

After Haley expanded into northwestern Colorado, he fenced his Wyoming ranches and bred his cows to purebred Herefords. Then his young beeves were moved onto his Colorado range for fattening. Later he fattened his cattle on sugar beet pulp produced near Sterling, Colorado.

Haley located his Colorado Two Bar headquarters on the Little Snake River. In time he grew to the northwest, building another ranch in Brown's Park on the Green River. At Wyoming's Rock Creek, Haley bred and raised horses on his Heart Ranch (his horse brand was a heart). He produced everything from work horses to polo ponies, and Two Bar cowboys were unusually well mounted. A Two Bar cowboy was assigned sixteen horses in two strings. Each Two Bar remuda was changed out at intervals of four to six weeks, providing a cowboy with eight fresh mounts, while the eight he had been working were turned out to pasture.

Capable Haley employees were paid generously and expected to assume considerable responsibility with little supervision from the owner. In Wyoming, black cowboy Thornt Biggs was a top hand who taught a generation of future wagon bosses and range managers the fine points of handling cattle. Young Willis Rankin hired on in 1877 and spent his entire career with Haley, becoming his right-hand man and troubleshooter. Haley's best employee in Colorado was the capable, experienced Hi Bernard, who produced great profits for years — until he fell in love with Haley's greatest nemesis, Ann Bassett.

Born in Texas in 1857, Hi Bernard began cowboying at the age of twelve, and at fifteen he helped shove a herd to Cheyenne. Two years later, Charles Goodnight put the impressive young drover in charge of a northbound herd, and this time Hi stayed in Wyoming. He worked for a number of outfits, including Ora Haley's in 1889. Five years later, Haley hired Bernard to buy and sell herds to take advantage of market fluctuations. By 1896 Bernard had been given complete authority to run Haley's Colorado ranches. During eight years as Colorado manager, associates estimated that he earned a million dollars for Haley — and they wondered why this astute cattleman did not accumulate any property for himself.

"I was paid a good salary, with all my expenses added," explained Bernard. "The company furnished me with a checkbook, and gave me a free rein to draw on company funds in connection with any business transaction I thought best for their interests, and they never questioned my judgment. . . . I was independent and my work suited me. I understood cattle and liked to work with them, and wanted to remain free from financial worries. I lived well, kept comfortable quarters at the ranches, and put up at the best hotels when I went to the cities."

But this arrangement, ideal for both Bernard and Haley, ended abruptly in 1904 when Hi married. His wife, Ann Bassett, hated Haley, spitefully opposing the cattle baron in every way possible. When she became Bernard's wife, Haley immediately severed relations with his most valuable employee.

Haley replaced Bernard with Heck Lytton, an experienced cattleman from Texas whose effectiveness was hampered by an incurably sour disposition. The next two managers lasted just six months each, before forty-year-old Bill Patton was placed in charge of Colorado operations in 1906. Although tough and hard-driving, Patton apparently ran the Two Bar more for his personal financial advantage than for Haley's. The terrible winter of 1908-09 cost the Two Bar severe cattle losses, and the image of the taciturn Haley was badly damaged by lengthy legal difficulties with Ann Bassett.

ANN BASSETT VS. ORA HALEY

Ora Haley was instrumental in organizing the ranchers of northwestern Colorado against rustlers. Two Bar manager Hi Bernard was a member of the first cattlemen's committee. Soon Tom Horn, operating under an alias, was collecting evidence against rustlers in Brown's Park. In separate incidents in 1900, a concealed rifleman assumed to have been Horn killed suspected stock thieves Matt Rash and Isom Dart.

Ann Bassett, the first white child born in Brown's Park (in 1878), was a neighbor and sometime sweetheart of Matt Rash. Pretty but hot-tempered, Ann went into a frenzy of anger against Ora Haley, whom she blamed for causing Rash's death. She elevated her off-and-on relationship with Rash to a tragic lost love, and obsessively regarded Haley as an overbearing range hog. For two years she hunted Two Bar cattle with diabolical intent.

"For hours on end we did nothing but 'jerk' Two Bar steers," related a companion who hero-worshiped the older Ann Bassett. "Riding full-tilt, we dropped the loops of our lariats over their rumps, flipping them in the air . . . when they came down, sometimes they broke their necks and sometimes they didn't." These cruel young women also forced small bunches of Two Bar cattle into the swift currents of the Green River, drowning some and scattering others into the badlands. In 1902 and 1903 they cost Haley "hundreds" of cattle.

Ann next took aim at Haley's gifted Colorado manager, Hi Bernard. It was easy for the attractive twenty-six-year-old to enthrall the forty-six-year-old bachelor, and they were married on April 13, 1904. Haley promptly discharged Bernard, and Hi moved onto his wife's little spread in Brown's Park. After that the marriage — having accomplished Ann's purpose — went downhill. (Hi would die in 1924.)

In 1911 Ann was tried for the theft of Two Bar cattle. Hi Bernard's testimony on behalf of Ann helped produce a hung jury. By the time Ann went to trial again, in 1912, she had divorced Hi. At her second trial a friendly jury ignored overwhelming evidence of her guilt. The entire crowd was hostile to Haley, and Ann was acquitted.

A victory dance at Craig's best hotel lasted until dawn. Haley's adversary lived until 1956, passing away at the age of seventy-eight. Only the good . . .

In 1914 Haley sold the Two Bar headquarters ranch on the Little Snake River to the Clay Springs Cattle Company of Fresno, California, for a reported $500,000. Rock Springs banker August Kendall bought the Two Bar spread on the Green River in Brown's Park. Four years later, a stroke confined Haley to a wheelchair, and he died in 1919 at the age of seventy-four. The longtime cattle baron had sold all of his livestock and ranches, and he left an estate of $2 million.

WHITE HOUSE RANCH HISTORIC SITE

In 1867 Walter Galloway homesteaded a 160-acre claim two miles northwest of Colorado "City" (the population was all of 300). He built a small cabin and worked his land, but found it necessary to hire out for wages in Colorado City. After seven years of struggle, Galloway sold out to Robert and Elsie Chambers, who intended to develop a truck farm. They built a handsome home from rock quarried on their property, as well as two steam-heated greenhouses and a big barn. The Rock Ledge Ranch became one of the most productive agricultural properties in El Paso County.

In 1900 the Rock Ledge Ranch was purchased by Gen. William Jackson Palmer, the founder of Colorado Springs. The Rock Ledge Ranch became a showplace of Palmer's estate, and in 1907 he erected a splendid house for his relatives, William and Charlotte Sclater. A prominent ornithologist, Sclater had been a museum director in South Africa, and he moved to Colorado Springs as director of the Colorado College Museum. Palmer had their house designed with South African Dutch features which would make the Sclaters comfortable. When this striking home was painted white, the site became known locally as the White House Ranch.

Today the White House Ranch Historic Site is a living history museum depicting agricultural life in the Pike's Peak region from 1860 to 1910. Attired in period dress, interpreters guide visitors through an 1860s homestead, the Rock Ledge Ranch of the 1890s, and the White House Ranch of the early twentieth century.

Built in 1907, this striking building inspired the name "White House Ranch."

Barn at the White House Historic Site near the Garden of the Gods.

Built in 1894, this handsome stone house was the home of Robert and Elsie Chambers, who developed the Rock Ledge Ranch.

COLORADO
RANCHING IN A CORRAL

Biggest Ranching Empires:
Prairie Cattle Company (2,240,000 acres; 59,000 cattle. The British syndicate ran another 90,000 head on 2,500,000 acres in northern New Mexico and the Oklahoma Panhandle); John Iliff's ranches (100 miles x 60 miles; 35,000 cattle).

Most Impressive Ranch Headquarters:
Two Circle Bar, in northwestern Colorado, owned by John, Robert and Sam Cary. In 1901 the Cary brothers hired 30 carpenters from Denver to build three two-story ranch houses (one for each brother and his family), along with barns, corrals and a two-story bunkhouse.

First Herd Trailed Into Colorado:
By John C. Dawson, in 1859 from Idaho Territory to Denver.

First Herd Shipped From Colorado:
By George Thompson, in 1869 from Las Animas to railroad cars in Kit Carson.

Best Bronc Stomper:
Jim Robinson, expert horse breaker from Craig.

Most Famous Ranch Animal:
Gentle the Twelfth, Hereford cow bought by John Prowers.

Best Rustlers' Haunt:
Brown's Park

Most Famous Rustler Killings:
Matt Rash and Isom Dart, in Brown's Park in 1900 by Tom Horn.

RANCHING IN WYOMING

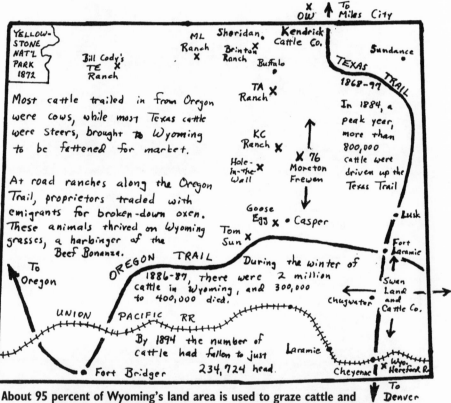

About 95 percent of Wyoming's land area is used to graze cattle and sheep. Cattlemen perpetrated many acts of violence against sheep-herders, but by 1894 sheep outnumbered cattle in Wyoming. By 1909 there were 4.5 million sheep in Wyoming, but thereafter these numbers declined, and today cattle provide the primary income from livestock.

At the Flag Ranch, near Laramie, stood a magnificent log structure called "The Big House" or "Homer Castle." It was built by Robert H. Homer in 1891 and burned in 1933.

— Courtesy Wyoming State Museum

The Pratt Ferris Ranch, east of Torrington, became the site of the first Red Cloud Agency.

— Courtesy Wyoming State Museum

Wrangler holding the horse of Rattle Snake Jack, the wolf trapper for the Big Red Ranch.

—Courtesy Wyoming State Museum

The ranch of Tom Sturgis on Horse Creek, 1884.

— Courtesy Wyoming State Museum

Cowboys gathered at Wyoming's Big Red Ranch in 1898.
— Courtesy Wyoming State Museum

7

Wyoming

O&W   KENDRICK CATTLE COMPANY

The Kendrick Cattle Company was built by tall, rugged John Kendrick, who worked his way from cowboy to cattle king, and whose superior leadership qualities then vaulted him to high political office. Kendrick was born in 1857 in Cherokee County, Texas, but his parents died in the 1860s. John and his sister were shuttled from one relative to another, while picking up a few grades of schooling — enough to give John a love of reading and a thirst for more knowledge.

At the age of sixteen John became a cowboy, soon hiring on with Charles Wulfjen, who had come to Texas from Wyoming. In 1879 Wulfjen sent John north with a herd to his ranch near Lusk in northeastern Wyoming, the ULA, named after Wulfjen's daughter Eula. The twenty-one-year-old Kendrick stayed on, working his way up to foreman. Sober, hard-working and intelligent, Kendrick avoided drinking and gambling sprees in favor of saving his wages and engaging in a program of self-education. He studied grammar, history, arithmetic — and, of course, the cattle industry, including livestock diseases, water supply, and marketing. When Wulfjen sold out to the Converse Cattle Company (which branded OW), Kendrick was hired as manager at $2,500 per year.

Already Kendrick had begun to purchase cattle with his savings, and he sold 860 head at a good price in September 1886 — just before winter blizzards brought on the Big Die-Up. After viewing the devastation wrought by overgrazing on the open range, Kendrick vowed: "I'll spend as much as on any cow (steer or calf) as that cow is worth."

On January 20, 1891, in Greeley, Colorado, the thirty-three-year-old Kendrick married seventeen-year-old Eula Wulfjen. Kendrick moved his bride into a five-room log house on the OW Ranch headquarters, across the Montana line about fifty miles northeast of Sheridan. Six years later the couple had a daughter, Rosa Maye, and three years after that a son, Manville. Eula taught her children the Three R's, and they learned to ride and shoot on the ranch. Manville later attended Phillips-Exeter Academy in New Hampshire, graduated from Harvard, then attended Ames Agricultural College in Iowa to prepare him for management of the family ranches.

THE OW ROUNDUP OF 1884

In April 1884 the executive committee of the powerful Wyoming Stock Growers Association planned roundups in thirty-one districts across the territory. Roundup No. 15, originating at the OW, would prove to be one of the largest in Wyoming history.

On May 15, 1884, nearly 200 riders — including John Kendrick — representing more than twenty outfits, gathered at the OW Ranch. There were 2,000 saddle horses, as well as numerous chuckwagons. The OW Roundup of '84 was assigned to scour the ranges of the Cheyenne River and its tributaries west of Hot Springs, South Dakota. Lasting six weeks, this massive operation rounded up and worked 400,000 cattle.

In 1897 Kendrick bought the property and livestock of the OW Ranch. So trustworthy was Kendrick that he appraised the property, counted the cattle and horses, then had his note accepted

John Kendrick's Trail End, along with its furnishings, cost $165,000.
— Courtesy Trail End Historic Center

for the price he named. During the late 1890s and early 1900s, Kendrick improved OW headquarters through the efforts of Swedish artisans bossed by burly Bylund Husman. The long cow shed was constructed of massive timbers stout enough to support a roof of flagstone slabs; there was an octagonal stone spring house, as well as a coal house, icehouse, stables and corrals; and the ranch house was built of huge, squared logs with dovetail notching at the corners. The logs were whitewashed for years, and the ranch home inevitably was called the "White House."

Beginning with the OW, Kendrick launched a program of resourceful land acquisition. He was one of the first ranchers to lease school lands from the states of Wyoming and Montana. Kendrick's purchases ranged from 40 to 6,000 acres. Working through a Minneapolis brokerage firm, he used land scrip issued by the federal government to veterans of the Civil War and the Spanish-American War. Also from the government he acquired land in the Ceded Strip on the Crow Reservation. He paid cash for homestead relinquishments, and once traded a quarter-ton REO Speed Wagon (with years of heavy duty on the OW) for a 640-acre tract on Remington Creek. Through such methods Kendrick put together a series of ranches in northeastern Wyoming and across the border in southern Montana. At these other ranches he constructed substantial buildings of logs and native stone, and industrious management of the Kendrick Cattle Company made him a millionaire.

The ballroom at Trail End covers the entire third floor. Note the orchestra balcony at right.

— Courtesy Trail End Historic Center

Note the carved buffet in the dining room at Trail End.

— Courtesy Trail End Historic Center

John Kendrick's favorite room at Trail End was his library.
— Courtesy Trail End Historic Center

KENDRICK CATTLE COMPANY PERSONNEL

Although ranch hands were required to address the boss as "Mr. Kendrick," John took a strong personal interest in his employees. Personnel with lengthy tenures and key roles in the Kendrick Cattle Company included:

Tug Wilson, who rode up the trail from Texas with Kendrick on an 1884 cattle drive, and worked at the OW until he died in 1934. He was wagon boss.

Clarence Wulfjen, Kendrick's brother-in-law, and a charter employee. When Kendrick died in 1933, Clarence managed the company until he retired four years later, stepping aside for Manville Kendrick.

Tom Reavis, Kendrick's cousin, came up from Texas to work at the OW just before 1900. He stayed for over sixty years, serving as foreman for more than forty years.

Tom Kendrick, a half-brother of John, came up from Texas in 1902, working mostly at the company's LX Ranch (he was foreman until his death in 1951).

Harry Davis, cowboyed on the LX from 1926 to 1958.

Chris Gunnell, cowboyed on the LX and OW from 1917 to 1957. He had a hot temper, and was imprisoned over a killing. But Governor John Kendrick pardoned him, then rehired him as a cowboy.

Jess Thomas, a fiercely loyal cowboy and wagon boss, 1929-1952.

Al Visborg, worked for the company forty years, 1946 to 1986, becoming a foreman at the age of thirty in 1959.

Spud Murphy, a highly efficient company manager, 1966-1985.

Harry Fuller, cowboy during the 1920s and 1930s. Acquired a "mail-order bride," but after delivering three daughters she ran off with a traveling salesman. Mrs. Tom Kendrick and fellow cowboys helped Harry raise the girls on the ranch.

Sam Miller, a homesteader hired during the 1930s to repair windmills. An artist with a copper still, he provided moonshine whiskey by the gallon to appreciative Kendrick cowboys.

For years Kendrick dreamed of building a home — his "trail's end" — on a magnificent site overlooking Sheridan from the west. Construction of Trail End began in 1908, but the three-story, eighteen-room mansion was not completed until 1913.

By that time Kendrick was heavily involved in politics. Virtually drafted into the Wyoming State Senate in 1910, he was elected governor in 1914, and the family moved to Cheyenne. Two years later Kendrick was elected to the United States Senate. Although a Democrat, Kendrick cooperated easily with the senior senator from Wyoming, Republican Francis W. Warren, who was a former governor and a millionaire woolgrower. While trying to protect the interests of his constituents, Senator Kendrick began inquiries that would uncover the notorious Teapot Dome scandal of the 1920s. He also found time to serve as president of the Wyoming Stock Growers Association and of the American National Live Stock Association.

In 1933, while in Sheridan, Kendrick died suddenly of a cerebral hemorrhage. At seventy-six he was the oldest man in the U.S. Senate. Eula Kendrick lived at Trail End until her death in 1961. Manville Kendrick operated the Kendrick Cattle Company with the same personal committment as his father. For a few years during the 1970s he stepped aside in favor of his son, John Kendrick II, but when John retired in 1978, Manville came out of retirement to resume control of the company. With no family member interested in succeeding him, Manville soon decided to sell off the company properties. With the sales in progress, Manville again retired, at the age of eighty-five in 1985. James Geurcio purchased the OW home ranch and handsomely restored the sturdy old buildings, while Trail End today is operated as a museum.

// THE SWAN TWO BAR

The Two Bar was the centerpiece of a cattle empire established by the legendary Alexander Hamilton Swan. Born in Pennsylvania in 1831, as a young man Swan moved to Iowa and spent nearly a decade and a half as a farmer and stockraiser. Gravitating to Wyoming during the post-Civil War cattle boom, in 1873 he organized the Swan Brothers Cattle Company with brothers Thomas and Henry and a nephew, Will Swan. Alex Swan acquired a herd on Chugwater Creek, where he established the Two Bar. He started another ranch on Sybille Creek to the west, then began using all of the open range in between. Although Alex pulled out of the Swan Brothers Cattle Company in 1880, he busily organized several other companies.

Swan's most ambitious promotion came in 1883, at the height of British investment fever over the "Beef Bonanza." Swan needed large amounts of capital for further expansion, but the cost of borrowing money in the United States was prohibitive. Persuasively, he extracted $2,387,685 from a Scottish investment group, forming the Swan Land and Cattle Company with headquarters in Edinburgh. Swan would be a stockholder and would receive a handsome salary as manager of ranching operations. In order to attract investors, western cattlemen usually overestimated ranching conditions and the size of their herds. "In our business we are often compelled to do certain things, which, to the inexperienced, seem a little crooked," admitted Swan.

Swan Land and Cattle Company riders, at the stockyards in Chugwater. Jimmie Danks is mounted on Charlie Boy, and Homer Payne stands beside him. Left to right: Delmer Wood, Ben Smith, Willis Troyer, Wally McVail, Johnny Robb (behind McVail), Lou Brundage, and night hawk John Brown (mugging Brundage).

— Courtesy Wyoming State Museum

A Swan Land and Cattle Company horse cavvy at a camp on Richeau Creek, ca. 1902.

— Courtesy Wyoming State Museum

Cashier George Milne snapped this revealing photograph of Swan Land and Cattle Company cowboys after a swim in Geyser Lake.

— Courtesy Wyoming State Museum

George Milne photographed this branding scene at the Chugwater stockyards, ca. 1900. Left to right: Billy Wilde, who became the last Two Bar foreman in 1903, Eddy Johnson, Phil Gatch, Delmer Wood, and Clayton Danks.

— Courtesy Wyoming State Museum

Two Bar headquarters at Chugwater. The cashier's office is at left.

Manager's residence, just north of the Two Bar office.

Warehouse at Two Bar headquarters.

Barn at Two Bar headquarters.

The Swan Land and Cattle Company acquired three additional ranches which provided the outfit with control of the entire range along Chugwater and Seville creeks. Greater expansion came in 1884, when the company contracted to pay $2.3 million for 550,000 acres of Union Pacific Railroad land to the west. Railroad lands were awarded in sections which alternated with sections of public lands, thereby giving Swan control of more than a million acres. Alex Swan established a Hereford breeding ranch near Cheyenne, and there was other land expansion. Soon the Swan Land and Cattle Company stretched from Fort Fred Steele, Wyoming, all the way east to Ogallala, Nebraska, and from Union Pacific tracks south to the Platte River.

In order to stock these vast ranges, cattle were purchased so aggressively that Swan foremen were issued company brand books. But the company's dominant brand remained the Two Bar. A complex of substantial buildings, begun in 1876, was erected along the Chugwater, and the Two Bar would remain the central ranch of the Swan Land and Cattle Company for nearly seven decades.

As Swan herds swelled to a reported 113,000 head, company stockholders were elated to receive dividends which averaged twenty-five percent in 1883, 1884, and 1885. The company employed 200 men, paying as much as $45 per month. Swan provided his cowboys with nine horses in each string, and the Two Bar gained a reputation for good chuck, specializing in desserts of rice-and-raisin pudding.

But the company's herd expansion was so rapid that even the enormous Swan ranges became overstocked, and the devastating winter of 1886-87 killed thousands of cattle. A tally revealed only 57,000 surviving cattle, and stockholders began to suspect that earlier book tallies had been greatly inflated. In May 1887 the Swan Land and Cattle Company went into receivership, a strong indication that the open-range cattle industry was ending.

Alex Swan, one of the founders of the Wyoming Stock Growers Association, a man of immense personal influence in Wyoming and in the western cattle industry, was discharged as manager of the ranch he had created. In the subsequent bankruptcy proceedings he lost his personal fortune and was sued by the company directors. Swan left Wyoming, tried to start over in Utah, but died in 1905 a broken man.

WYOMING'S FIRST RODEO

Shortly after organizing the Swan Land and Cattle Company, Alex Swan staged Wyoming's first rodeo as an entertainment for the British investors. About 150 of these owners journeyed to Cheyenne, where Swan met them with carriages, wagons, and saddle horses. The entourage headed out to the Laramie Plains, where a barbeque was laid out for them, and where every rider on the payroll participated in the Wild West exhibition.

Horse racing, roping, and bareback riding, of broncs and steers, amused the stockholders and other visitors who had followed the crowd out of town. Swan had brought Indians to add to the western atmosphere, and a tug of war was staged between cowboys and Indians. Teenaged cowboy Robert LeRoy Parker, who later earned notoriety as outlaw Butch Cassidy, demonstrated his marksmanship with a revolver.

The most unique event placed twenty-five cowboys atop broncs that had never been ridden. Blindfolded and placed in a line, the horses were supposed to race 200 yards. But when the blindfolds were pulled, the broncs began bucking and many of the riders were thrown. It was half an hour before one skillful rider managed to guide his raw mount across the finish line.

The stockholders spent two nights under the stars. Many suffered insect bites and sunburn, but Swan was informed that his spectacle was worth coming 6,000 miles to see.

Other small rodeos soon began to be held throughout Wyoming, and in 1897 the first Cheyenne Frontier Days opened the annual run of one of the greatest of all rodeos. And in 1896 the most famous of the early rodeo broncs was foaled on the Two Bar at Chugwater. For fifteen years Steamboat appeared at the most prestigious rodeos, including Frontier Days each year, and threw almost every rider who mounted him. The fierce black bronc from the Two Bar earned the title "King of the Plains."

Swan was replaced as Two Bar manager by John Clay, who had been involved with the financial trust that had controlled the company. Clay was rumored to have hired Tom Horn as a range detective, and he strongly supported the Regulators during the Johnson County War of 1892. But in the wake of public disapproval of the high-handed invasion of Johnson County, Clay was dismissed by the directors.

As conditions in the range cattle industry worsened, company directors reduced operations efficiently and continued ranching. An eventual reorganization shifted the board to a group of American directors. In 1904 the famous Swan Land and Cattle Company switched to sheep, a sensible move made by many cattle ranches, because sheep produce two annual payoffs, their fleece and the lamb crop.

The company continued to sell off its lands, until by 1948 only 2,183 acres of the Two Bar headquarters remained. The company was officially dissolved on December 20, 1951, and the remains of the Two Bar were given to manager Curtis Templin, who had served the ranch for thirty-five years. Today the old Two Bar headquarters buildings stand deserted and locked, but remain easily accessible to travelers who enjoy a nostalgic visit to the nerve center of one of the largest ranches of the western frontier.

Standing in Chugwater is a Dempster, the most popular windmill in the early 1900s.

WHR WYOMING HEREFORD RANCH

The world-famous Wyoming Hereford Ranch was originated in 1883 by cattle king Alexander Swan, who was determined to introduce Herefords at a time when Shorthorns dominated breeding in Wyoming. In conjunction with his cattle business, Swan was vice-president of the First National Bank of Cheyenne, and he located his Hereford breeding ranch six miles southeast of Wyoming's capitol. Crow Creek meandered through the ranch, and adequate precipitation produced fine grasses. An elevation exceeding 6,000 feet enhanced the quality of the grasses.

Swan sent an English-born associate, George F. Morgan, to England to purchase foundation stock. Morgan brought back 146 Herefords, including a 2,600-pound bull, Rudolph. Although Rudolph died in 1885 of injuries incurred while being loaded, he produced eighteen sons and twenty daughters during his brief tenure, and his commanding frame and 1884 show prizes went far in spreading acceptance of Herefords. Morgan helped form the American Hereford Cattle Breeders Association, and Swan would become its fifth president.

But the conditions which wrecked the open-range cattle industry after 1886 thrust Swan into bankruptcy, in 1887. The Hereford Ranch went into receivership, and the appointed receiver was Colin Hunter, who intended to preserve the herd and ranch intact. Will Rossman, trained by George Morgan, was hired as foreman. Hunter finally sold the outfit in 1890, to Cheyenne businessman Henry Altman and his partner, Dan McUlvan, who retained Rossman. "The best today, not good enough for tomorrow," was the motto of the progressive Altman, who operated the Wyoming Hereford Ranch for more than a quarter of a century.

In 1916, when Altman was seventy-two and beef prices were accelerating because of wartime demands, he and McUlvan sold the ranch to Raymond S. Husted of Denver. The purchase price was $400,000: $10,000 was paid to bind the deal; possession would be given when an installment of $150,000 was delivered; and the balance was to be paid in installments within five years. Rossman was to stay on as superintendent, and the transaction included 18,968

The fine old horse barn of the Wyoming Hereford Ranch.

Interior of the horse barn.

acres and the "World's Greatest Herd of Registered Herefords." But drought and a precipitous postwar price decline led to receivership in 1921.

Again the herd and ranch were sold intact, to a group headed by Henry Parsons Crowell of Chicago, founder of the Quaker Oats Company. Rossman continued as superintendent until he retired in 1927, after forty-three years of service to the ranch. By this time the ranch had entered the era of a remarkable bull, Prince Domino (1914-1930). The American Hereford Association's 1981 "Genealogical Listing of Hereford Sires" devoted seventy-nine of 100 pages to sires descended from Prince Domino, and a prominent marker designates his grave on the ranch.

At the age of eighty-three in 1938, Crowell stepped down from active leadership of the ranch, but continuity was provided by director and longtime manager Robert Lazear. Shortly before his death in 1957, Lazear engineered sale of the ranch to experienced Hereford ranchers G. C. Parker and Mr. and Mrs. T. E. Leavey of California. At this point the Wyoming Hereford Ranch consisted of 2,500 purebred cattle and 57,000 deeded and leased acres, and the purchase price exceeded $2 million.

The bunkhouse, remodeled since its original construction, but unused in recent years.

In 1967 the ranch was bought by Nielson Enterprises, Inc. of Cody. The next year a prairie fire raced through knee-high grass so rapidly that ninety-three head of cattle were badly burned. Three days later, a blizzard hampered treatment, and forty-three of the injured animals had to be destroyed. A rebuilding program was begun, but within a few years the cattle industry entered a severe slump.

The herd was auctioned off in 1976, and two years later the ranch was sold in fourteen parcels. The largest parcel, which included the ranch buildings and 11,300 acres, was acquired by Dr. Sloan Hales and his wife, Anna Marie. Under their ownership the ranch continued its historic focus on registered Herefords.

Today the Wyoming Hereford Ranch is a major attraction of the Cheyenne area, as visitors enjoy self-guided tours. Group tours, picnics, campouts, and hayrides may be scheduled, and each year the ranch hosts the Cheyenne Symphony, a fundraiser for the Cheyenne Symphony Orchestra. Well into its second century of operation, the Wyoming Hereford Ranch is a tangible link with the ranching heritage of the Old West.

A recent product of the Wyoming Hereford Ranch.

76 MORETON FREWEN'S 76 RANCH

"I begin to believe that being a gentleman is much against one here, as in the colonies generally." The Englishman who came to his conclusion about the "colonies" was visionary cattleman Moreton Frewen, who confided to his wife that "these western men, at least some of them, are the most impracticable, aggressively independent people possible, and I often long to thrash one or two but for the want of dignity in such a proceeding."

Frewen, a proper English gentleman, inevitably became the butt of humor by Wyoming cowboys. Their favorite story was that he bought his first herd twice, after they cleverly drove the cattle around a hill to be tallied a second time by the greenhorn foreigner. Although proof exists that this incident never happened, the image of everyday drovers outfoxing an English dude was irresistible. Frewen, however, invested his entire personal fortune in the West, and he displayed many of the finest qualities of frontiersmen, including courage, stamina, and boundless optimism.

In 1878, when he was twenty-five, Frewen visited the JAs Ranch owned by a personal friend, John Adair, and Charles Goodnight. Captivated by the West and by the prospect of immense profits available through open-range ranching, Frewen headed north to find his own unclaimed rangelands. Riding in Wyoming with guides and his older brother, Dick, in December 1878, Frewen and his party headed their tiring mounts into a path broken through deep snow by a buffalo herd. At the Powder River Valley, Moreton Frewen determined to build a cattle empire in that magnificent but isolated region.

Moreton had $39,000 from an inheritance, which he pooled with Dick's larger amount, but he intended to raise far more capital through English investors. Partially to impress prospective investors, he built a two-story log cabin, sixty feet on a side, modeled after British hunting lodges. Erected in 1879, "Frewen's Castle," as frontiersmen dubbed it, stood a hundred yards back from the Powder River, just below the junction of the North and Middle Forks. Although the big log structure was 200 miles north of the nearest railroad station, Frewen imported fine hardwoods from England for the interior, while furnishings, windows, doors, and shingles from Chicago were hauled by ox-teams from the railroad. The main room was forty feet square, with two big fireplaces and a wide walnut stairway leading up to a mezzanine balcony, where an orchestra soon would play for dinner guests and dancers. Also on the first floor were sitting rooms, an office, library, kitchen and pantry; bedrooms were on the second floor. Stables, corrals, bunkhouses, and other outbuildings were built near Frewen's big lodge.

Frewen bought 2,000 head of Shorthorns bearing a 76 brand, and newspapers reported that also in 1879 he purchased 10,000 cattle from Montana and 7,000 from Texas. Fred G. S. Hesse, a young, capable Englishman, was hired as foreman at $3,000 per year, and in time he would run his own ranching operation.

Frewen wrote countless letters inviting wealthy Englishmen to come to his ranch to hunt and play polo and consider investment prospects. Guests were escorted from the railroad to carriages, and were met at the 76 by butlers, maids, valets, and chefs. They were conducted into the Big Horn Mountains for hunting aboard quality mounts, with wagons hauling camp supplies, tents, beds, and portable bathtubs.

The enthusiastic Frewen persuaded most of his well-heeled visitors to help him form the Powder River Cattle Company. Frewen agreed to serve as manager for no salary, although he would enjoy a large portion of any profits. During ensuing years he would travel to England frequently, and he bought a small house in Cheyenne, where the company kept an office run by Englishman Frank Kemp. But Dick Frewen had helped to organize the prestigious Cheyenne Club, which Moreton soon joined, and where he often dined and slept.

Moreton was six-foot-three, blond, mustachioed, and handsome. With utter confidence he asserted that every woman he had ever bedded had been "completely paralyzed by the vigour" of his physical charms. While in New York early in 1881, he met thirty-one-year-old Clara Jerome, the eldest of three daughters of millionaire Leonard Jerome, and they wed in April 1881. (All three Jerome daughters married Britishers. Jennie Jerome became the mother of Winston Churchill in 1874.) Despite a guest list that bristled with English nobility and U.S. Army officers and their wives, Frewen's bride was daunted by the isolation of the ranch. After suffering a miscarriage she moved back to the more civilized environs of New York.

Moreton visited New York as often as possible, usually on his way to or from England, but once he received the infusion of Powder River Cattle Company capital, Frewen was busier then ever in Wyoming. In the spring of 1883, 9,824 calves were branded with the 76. During that year Frewen purchased 50,000 head of cattle from Nebraska and Oregon, and he imported fifty purebred Sussex bulls. Soon it was estimated that 70,000 cattle grazed Frewen's

THE CHEYENNE CLUB

Englishmen and Easterners who ventured west to engage in cattle ranching brought with them an aura of sophistication which appealed to successful Westerners desirous of displaying the affluence they had earned on the frontier. Western ranchers who had carved cattle empires from a primitive land wanted to prove that they could be gentlemen, like Englishmen and Easterners. Soon gentlemen's clubs, which abounded in England and in the East, began to be organized across the West. Among the better clubs were the Denver Club and the Montana Club, in Helena. Most notable of all, however, was the Cheyenne Club, chartered in 1880. Dick Frewen was one of the twelve founders, and his brother, Moreton, became an early member.

The club originally admitted fifty members, but demand for entry to this fashionable establishment quickly raised the membership level to two hundred, with an initiation fee of $100. Applicants had to be sponsored by a club member, and even after a careful background check an aspiring applicant might be blackballed. Intoxication and fighting were forbidden, and so were profanity and cheating at cards. When one of the founders, C. M. Oelrichs, drunkenly struck a Cheyenne Club bartender, his membership was rescinded.

This exclusive group was housed in a three-story frame and brick building with a mansard roof and a sweeping veranda. There was a library which boasted the leading newspapers and magazines, a billiard room, two wine vaults, and a dumbwaiter which sent gourmet meals from the basement kitchen up to the wood-paneled dining room. Members played chess on the veranda and tennis on the Cheyenne Club court, and staged harness races and polo matches. "No wonder they like the club at Cheyenne," approved the cultivated and widely traveled Owen Wister, who belonged to a club in Philadelphia. "It's the pearl of the prairies."

The Cheyenne Club.
— Courtesy Wyoming State Archives

Cheyenne Club dining room.
— Courtesy Wyoming State Archives

range, from the Tongue River south to Teapot Dome, and from Hole-in-the-Wall country east to the Black Hills.

Frewen established feeding pens in the Nebraska corn belt, and built a slaughter house at Superior, Wisconsin, so that he could send dressed beef on refrigerated ships across the Great Lakes to England. He erected a packing plant atop Sherman Hill, the highest point on the Union Pacific, between Laramie and Cheyenne, utilizing the frigid temperatures to keep his beef frozen. Twenty miles downriver from his ranch headquarters, Frewen built a post office and general store. Then he strung Wyoming's first telephone line north of the Platte, connecting his ranch with the post office, and from there with the outside world through a military telegraph line.

But Frewen recognized that the ranges had been rapidly overstocked and overgrazed. "I dread the coming winter," he stated prophetically in September 1886. "If it is a severe one half of the cattle in Wyoming will die for sure."

It was a severe one, of course, resulting in the Big Die-Up that effectively finished open-range ranching and the British Beef Bonanza. Foreseeing this collapse, Frewen went to Canada and arranged a long-term lease for 80,000 acres of grassland. But the board in faraway England refused to approve the move into Canada, part of a vigorous policy clash which resulted in Frewen's dismissal from the company he had created. The Powder River Cattle Company liquidated soon after the Big Die-Up, and the shareholders lost their investments.

Moreton Frewen never returned to Wyoming. With typical energy and enthusiasm he ventured to the corners of the world in pursuit of fortune, but none of his business schemes worked out. He died in 1924 at the age of seventy-one. Although Frewen's imaginative and large-scale approach to open-range ranching proved short-lived, survivors of his enormous 76 herds stocked the ranches of numerous small operators who swung a large loop.

TA   TA RANCH

Like many other western ranches, the TA originated during the Beef Bonanza. In 1882 Doc Harris became the first cattleman to use the TA brand, at a location thirteen miles south of Buffalo,

Wyoming. The TA never was especially large — about 8,000 acres were grazed at one time — and there were no famous individuals who owned the ranch. But the TA enjoyed an extraordinary moment in frontier history as the site of the climactic battle of the West's most notorious range war.

By 1892 members of the powerful Wyoming Stock Growers Association had decided to hire gunmen for an extralegal invasion of Johnson County in northern Wyoming. Members of the association were regarded as arrogant cattle barons, and many were wealthy absentee owners. Their immense herds were tempting targets for rustlers, who often homesteaded claims and started their own ranching operations with stolen cattle. There was a concentration of these rustler/ranchers in Johnson County, and they became even more defiant after two of their number, Ranger Jones and John Tisdale, were assassinated late in 1891, presumably by a hired gun of the association. Willing to take more drastic action, 100 association members subscribed $1,000 each for an "extermination fund." Saddle horses were purchased. along with three big wagons, guns

Log cabin on the TA.

and ammunition, tents, bedding, and miscellaneous supplies. "Regulators" were recruited in Texas with an offer of $5 per day, all expenses paid, a $50 bonus to each mercenary for every rustler killed, and a $3,000 insurance policy. Twenty-two Texas gunmen headed north and on April 5, 1892, boarded a special Pullman car in Denver. Later in the day the Pullman reached Cheyenne, where the Wyoming Stock Growers Association had just concluded the annual spring meeting. The Pullman was added to a train which included three stockcars for the horses, a flatcar for the three wagons, a baggage car, and a caboose. Nineteen of the sponsoring cattlemen joined the expedition, along with five of the association

TA stable interior, where horses under fire were taken for shelter.

Interior view of the hayloft above the stable shows bullet holes and loopholes from 1896.

The log stable at the TA. The wing at left was added after the 1896 battle.

"detectives" and two newspapermen. A list of seventy men to be shot or hanged included Sheriff Red Angus and his deputies.

The train reached Casper early on April 6, whereupon the Regulators debarked, mounted their horses, then headed for Buffalo more than 100 miles to the north. Telegraph lines to Buffalo were cut, and the Regulators rode northward with the intention of killing as many men as possible on the hit list and taking over law enforcement in Johnson County.

As the Regulators passed the Hole-in-the-Wall area, they learned that Nate Champion and Nick Ray, prominent on the death list, had leased the nearby KC Ranch. Deploying around the ranch before daybreak on Saturday, April 9, the Regulators besieged the little cabin throughout the day. Ray was fatally wounded with a surprise volley, but Champion courageously held out until late in the afternoon, when a fire wagon engulfed his cabin in flames. He made a break, but twenty-eight bullets ripped the life from his body.

The log TA ranch house was a primary target during the climactic battle of the Johnson County War.

The Regulators resumed their march, riding past the buildings of the TA Ranch. But early on Sunday morning, with the column halfway between the TA and Buffalo, a warning was received that Sheriff Red Angus was rousing the countryside against this invasion. It was decided to retreat and fort up at the TA. Doc Harris was friendly toward the Regulators, but he was away, and only a seventeen-year-old employee was present.

The TA buildings stood in a windbreak of trees along a bend of Crazy Woman Creek. The ranch house was built of heavy logs, and the complex also included a log stable, an icehouse, a dugout which served as a potato cellar, and a hen house. Timbers once cut for a building project now were used to erect a twelve-by-fourteen redoubt atop a hill, while the ranch house was ringed with breastworks.

At dawn on Monday, April 11, the Regulators could see that at least fifty men had dug earthworks on surrounding hills during the night. Soon Sheriff Angus arrived with forty reinforcements, then returned to Buffalo to organize more armed men. By the end of the day 300 angry citizens encircled the TA, while women prepared food behind the lines.

When the attackers opened fire, they targeted the saddle horses in the corral, until a few of the Regulators dashed out to lead the survivors into the adjacent stable. Then the besiegers zeroed in on the ranch house, which became, according to one Regulator, "considerably shot up." Amazingly, one of the cattlemen retired to a bedroom to take his regular afternoon nap during the siege.

The Regulators held out Monday and Tuesday, but on Wednesday morning they were dismayed to see a portable breastwork of timbers and hay bales slowly moving toward the ranch. This siege engine was rolling on the wheels of the Regulators' supply wagons, which had been loaded with guns, ammunition — and dynamite. A heavy fire was directed against the device, but with no effect. When the breastwork moved close enough, the citizens would be able to lob sticks of dynamite at the ranch buildings.

Suddenly, in best movie fashion, three troops of cavalry rode to the rescue. The Regulators thought that now they would have to take on the army, but orders from Washington to Fort McKinney, west of Buffalo, directed that the cattlemen and their mercenaries should be protected. On Tuesday a courier had slipped past the besiegers and delivered a message for help to influential cattlemen, who contacted the governor and both Wyoming senators. The senators had awakened President Benjamin Harrison at the White House, and telegraphed instructions soon were dispatched to Fort

McKinney. The Regulators surrendered and were taken to the fort, safe from lynch mobs.

"They were mine," complained an exasperated Red Angus. "I had them in my grasp and they were taken from me."

The only casualties were two Texans who accidentally, and fatally, wounded themselves. Taken to Fort Russell for trial in nearby Cheyenne, the Regulators enjoyed a raucous celebration in the Cheyenne Club when charges finally were dropped.

Following a wild four days in April 1892, the TA quietly continued to operate as a cattle ranch. Soon after the TA was used as a battlefield, Doc Harris traded his outfit to Clarence Gammon for the Lightning Flats Ranch in the Sundance area. The Gammon family ran the TA until the early 1980s, after which the spread was controlled by a cattle company, then a bank. In the 1990s Earl and Barb Madsen have extended the hospitality of their historic old ranch to guests, who may wander among sturdy buildings which still display century-old bullet holes.

The TA ranch house faces south, and has been refurbished for guests.

GOOSE EGG RANCH

The Searight Brothers Cattle Company originally operated in the Chugwater area, but the Searights sold out in 1878. The next year the brothers trailed a herd of cattle from Oregon to their new Goose Egg Ranch, twelve miles west of Casper. They located head-quarters on a flat where Poison Spider Creek flows into the North Platte River, and in the summer of 1879 a bunkhouse, barn, and other structures were erected.

In the fall of 1881 construction began on a big stone ranch house, which was completed the following spring. Goose Egg fore-man W. P. Ricketts and his cowboys dug the basement and quarried stone at a nearby outcrop. Lumber and other materials were brought in on freight wagons from Cheyenne, while stonemasons and a carpenter also were imported from Cheyenne.

The house faced east and the front door opened onto a hallway which gave access to the downstairs rooms and to a stairway which led to four upstairs bedrooms. Downstairs, to the left of the hall-way, was the living room, and in the rear, a bedroom; a big double fireplace heated both rooms. To the right of the hall were the din-ing room, kitchen, and two pantries, as well as a washroom at the side entrance.

The Goose Egg ranch house, made famous by Owen Wister in The Virginian, *originally boasted a porch across the front.*
—Courtesy Wyoming State Museum

In 1886 the Goose Egg was purchased by Joseph M. Carey, who owned extensive cattle oper-ations and who would serve Wyoming in numerous political ca-pacities, including governor and senator. Carey's Goose Egg fore-man was William Clark, who ap-parently operated the imposing stone house as a hotel for a time. During his Wyoming travels Owen Wister visited the Goose Egg, and in his landmark 1902 novel, *The Virginian*, the ranch was the setting for a dance. In one of the book's most memo-rable scenes, the Virginian and a friend sneak upstairs during the dance and switch sleeping babies, causing parents to leave with the wrong children.

Carey eventually sold the Goose Egg Ranch to Dan Speas, who attempted to have the deteriorating house preserved as a historical site. But souvenir hunters caused growing damage to the old house, and it was razed in 1951.

TOM SUN'S HUB AND SPOKE RANCH

When Tom Sun established his Hub and Spoke Ranch in the shadow of Oregon Trail landmark Devil's Gate, he became perhaps the first permanent settler in Wyoming's beautiful Sweetwater Valley.

Of French Canadian heritage, Thomas de Beau Soleil (later anglicized to Tom Sun) was born in Vermont in 1846. When Tom was eleven his mother died and he ran away from home. A trapper took the boy under his wing, and on a trapping trip to Wyoming Tom first visited his future home. A trading post had been erected in 1858 just south of Devil's Gate along the Sweetwater River, barely a hundred yards from the site of Tom's future ranch buildings.

During the Civil War, Tom Sun worked on railroad construction for the Union Army, and he continued such employment after the war as the Union Pacific built into Wyoming. Along with

Corrals at the Hub and Spoke Ranch, with Devil's Gate in the background.
— Photo by Jim Browning

Tom Sun's original ranch house, built of squared logs and whitewashed.
— Photo by Jim Browning

Tom Sun's ranch buildings, begun in the 1870s, have been placed on the National Register of Historic Places.
— Photo by Jim Browning

Buffalo Bill Cody, he was hired as a scout and guide out of Fort Fred Steele. In 1872 he moved to Devil's Gate, erecting a cabin and regularly adding to the buildings of his Hub and Spoke Ranch. He married in 1883, and reared four children at the Hub and Spoke.

One of Tom Sun's neighbors in the Sweetwater Valley was Jim Averell, a former army sergeant who established a road ranch and post office on his homestead. In 1886 Averell married Ellen Watson, but the marriage was kept secret so that she could file a separate claim in the Sweetwater Valley, and both properties were fenced. Cattleman Albert John Bothwell was especially resentful at being fenced off of a favorite hay meadow, and he tried to arouse the ire of other ranchers. Ellen had accumulated a small herd of cattle, and Bothwell called five neighboring stockmen together on his former hay meadow and convinced them, apparently with the lubricating assistance of liquor, that Jim and Ellen were rustlers.

On July 20, 1889, Bothwell and his friends, including Tom Sun, arrogantly inspected Ellen's cattle. Despite her protests that she could produce a bill of sale, she was roughly forced into the back seat of Tom Sun's buggy. As the ranchers approached Averell's road ranch, he was seized and placed beside her. Covered with rifles, Jim and Ellen were taken a considerable distance and crudely hanged, slowly strangling to death. There were several witnesses to the abduction. Frank Buchanan, a close friend of the victims, followed the procession and tried to break up the execution with revolver fire, until he was driven away by rifle bullets.

The cattlemen persuaded Ed Towse, a reporter for the influential *Cheyenne Daily Leader*, to depict Jim and Ellen as rustlers and general villains. While in the army Jim had fatally wounded a bullying civilian in a barroom brawl, and Towse described him as a "murderous coward" and cattle thief. Ellen was deliberately confused with two notorious Wyoming prostitutes, "Cattle Kate" Maxwell and Ella Watson. These and other misconceptions were challenged in competing newspapers, but the witnesses began to die or move away, and high-powered legal talent secured the release of the lynchers.

After being released from custody, Tom Sun returned to his beloved ranch. Cattlemen promptly elected him to the executive committee of the Wyoming Stock Growers Association, a position he would hold until he died. He worked as a hunting guide and as a prospector to supplement his ranch income, and he patiently acquired title to various tracts as homesteaders crowded into the valley. In order to qualify for 640 acres under the Desert Land Act, Sun built waterwheels on the Sweetwater River and on Cherry Creek to raise the water high enough to irrigate the land.

A fixture at the ranch was the Japanese cook, Bobby, who cooked rice for every meal, even though he was the only one who ever ate it. After many years of service, Bobby died and was buried in the Sun family plot in Rawlins. Another fixture was June, a pet antelope who regularly wandered into the cabin after supper and ate leftovers from plates on the table. A less welcome animal was a huge male wolf who dragged down a calf every night. Tom Sun, Jr., killed the beast, and the hide was tanned and today is displayed in the Sun Ranch Room in the Carbon County Museum.

Tom Sun continued to develop his ranch until his health failed. He died in Denver in 1909, but his twenty-five-year-old son, Tom Sun, Jr., assumed control of the Hub and Spoke. Under his direction a quality Hereford herd was built and ranch expansion continued. After Tom, Jr.'s death in 1975, the ranch founder's grandson,

Bernard Sun, maintained family ownership and management. The buildings put up more than a century ago by Tom Sun have been placed on the National Register of Historic Places, and history-minded tourists in the area will discover Devil's Gate, Independence Rock — and Tom Sun's ranch.

Tom Sun's hat, chaps, and saddle, on display in the Carbon County Museum in Rawlings.
    — Photo by Jim Browning

ℳ   MASON-LOVELL RANCH

Cattle were ranched primarily on the southeastern ranges of Wyoming until the mid-1870s, when hostile Indians finally were confined to reservations by the military. Ranchers penetrated the magnificent Big Horn Basin in northern Wyoming in 1879. Among these pioneer cattlemen was Henry Lovell, who was born in 1838 in Michigan. As a young man Lovell gravitated to the West, working for a time guarding government mail trains. His route ran from Fort Dodge to Mexico City, and in the course of his duties he collected three bullet wounds and a knife scar.

Lovell wisely left this work to learn the cattle business in Texas. Soon he sought open range to the north, establishing two

One of four abandoned log buildings still standing at ML headquarters.

ranches in the Big Horn Basin. Forming a partnership with Anthony Mason, a Kansas City capitalist, Lovell pushed a herd from Kansas to Wyoming in 1880. Lovell and Mason organized one of the first ranches on the east side of the Big Horn Basin, ranging their ML cattle from Pryor Gap all the way south to Thermopolis. A headquarters complex was built in 1883 on the banks of Willow Creek, about thirteen miles east of present-day Lovell. The ML herd size soon exceeded 25,000 head, but Mason died in 1892 and the estate settlement forced Lovell to reduce his operation.

Though Lovell finally liquidated the ranch in 1902, a post office continued out of ML headquarters from 1894 until 1908. Today the abandoned log buildings may be visited at a haunting site just south of U.S. Highway 14 Alternate.

ML Ranch headquarters, shaded by cottonwoods and sheltered by the mountains.

The ML blacksmith shop.

One of the bunkhouse dogtrots.

The sprawling ML bunkhouse is divided by two dogtrots.

BRADFORD BRINTON RANCH

William Moncreiffe migrated to the West from Scotland in 1885. Captivated by Wyoming, William and his brother, Malcolm, developed a ranch at the base of the massive Big Horn Mountains, just a few miles south of Big Horn. The brand was a Quarter Circle A, and in 1892 the brothers built a two-story frame house on the banks of Little Goose Creek. Perhaps their most notable transaction came during the Boer War at the turn of the century, when 2,500 cavalry and draft horses were sold to the British Army.

In 1923 the Quarter Circle A was purchased by Bradford Brinton, a businessman, military officer, sportsman, and patron of the arts from Illinois. Brinton enjoyed the life of a gentleman rancher on his 2,400 acres, and he enlarged his house to twenty rooms so that he could display his collection of western art, books, historical documents, and Indian artifacts.

Helen Brinton survived her brother and, wishing to share his fine acquisitions with future generations, established the Bradford Brinton Memorial Ranch Museum. Today tourists may enjoy the handsomely furnished house and outbuildings, as well as a visitor center where much of the collection is displayed.

Horse trough at the Brinton Ranch.

The Moncreiffe brothers built this house in 1892, but new owner Bradford Brinton made a major addition thirty years later.

Milk house at the Brinton Ranch.

The barn and other structures of the Brinton Ranch stand in the shadow of the Big Horn Mountains.

WYOMING
RANCHING IN A CORRAL

Biggest Ranches:
Swan Land and Cattle Company (1,500,00 acres; 113,000 cattle; 200 cowboys)
Moreton Frewen's 76 (70,000 cattle; 75 cowboys)

Best Cattle Towns:
Cheyenne, Laramie, Buffalo, Medicine Bow

Best Rodeo: Cheyenne Frontier Days, 1897-Present

Most Powerful Cattlemen's Organization:
Wyoming Stock Growers Association

Most Famous Cattlemen's Hired Killer:
Tom Horn

Best Outlaw Hideout:
Hole-in-the-Wall

Most Notorious Lynching of Accused Rustlers:
"Cattle Kate" Watson and Jim Averell, July 20, 1889

Best Ranch Battlefield (Tie):
KC Ranch (April 9, 1892) Nate Champion vs. Regulators
TA Ranch (April 11-13, 1892) Regulators vs. 300 citizens

Most Famous Range War:
Johnson County War, 1892
R.I.P. John Tisdale, Nick Ray, Nate and Dudley Champion

Most Influential Wyoming Cowboy Novel:
Owen Wister, *The Virginian,* 1902

Cowboys' Most Memorable Sight:
New Yorker Richard Trimble, partner in the Teschemaker & DeBillier Cattle Co., strolling through his herds with a pet poodle on a leash.

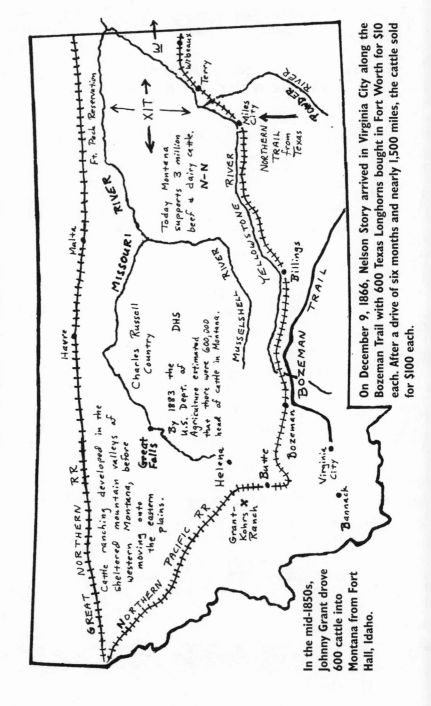

RANCHING IN MONTANA

GREAT NORTHERN RR

Havre

Malta

Ft. Peck Reservation

MISSOURI RIVER

Cattle ranching developed in the sheltered mountain valleys of western Montana, before moving onto the eastern plains.

Charles Russell Country

By 1883 the U.S. Dept. of Agriculture estimated that there were 600,000 head of cattle in Montana.

DHS

Great Falls

NORTHERN PACIFIC RR

Helena

Grant-Kohrs X Ranch

Butte

Bozeman

MUSSELSHELL RIVER

Today Montana supports 3 million beef & dairy cattle.

N-N

W

XIT

Wibeaux

Terry

POWDER RIVER

Miles City

NORTHERN TRAIL from Texas

YELLOWSTONE RIVER

Billings

BOZEMAN TRAIL

Virginia City

Bannack

In the mid-1850s, Johnny Grant drove 600 cattle into Montana from Fort Hall, Idaho.

On December 9, 1866, Nelson Story arrived in Virginia City along the Bozeman Trail with 600 Texas Longhorns bought in Fort Worth for $10 each. After a drive of six months and nearly 1,500 miles, the cattle sold for $100 each.

8

Montana

CK GRANT-KOHRS RANCH

During the early 1970s, the National Park Service initiated a search for a western ranch which could be developed into a living history museum depicting the cattle frontier. At Deer Lodge pioneer rancher Johnny Grant had founded Montana's first substantial spread, then sold it to Conrad Kohrs, who became famous as the "King of Montana Cattlemen." Deciding that no better example of a notable frontier ranch was available, the Park Service acquired core acreage and a well-preserved set of substantial buildings and corrals. Today visitors experience the nostalgic atmosphere of one of the most colorful chapters of western history at the Grant-Kohrs Ranch.

Johnny Grant was a Canadian trapper who arrived in 1855 at the grassy valley where he would build his ranch. Although he had agreed to manage the new Hudson's Bay Company trading post in the valley, Grant soon began developing his own operation. He built a long cabin of cottonwood logs on a gentle ridge east of the Clark Fork of the Columbia River. His wife, a full-blood Bannock named Quarra, kept a tepee in the yard where she preferred to sleep.

Deer Lodge teemed with deer, elk and other wildlife, and to the east loomed the snow-capped mountains of the Continental Divide. Although winter temperatures often plunged to subzero

213

As late as the 1930s the CK Ranch raised and sold big Belgian draft horses.
— Courtesy Grant-Kohrs Ranch

Longtime CK cook Ham Sam.
— Courtesy Grant-Kohrs Ranch

Registered Herefords grazing behind the CK Ranch house.
— Courtesy Grant-Kohrs Ranch

In 1890 Conrad Kohrs added a large brick wing to the rear of the original clapboard house.
— Courtesy Grant-Kohrs Ranch

levels, snow barely fell in the valley, which was about thirty miles north to south and seven miles wide. This sheltered valley was about to become the cradle of Montana's cattle industry.

When the town of Deer Lodge took shape south of his cabin, Grant began to trail in cattle from Idaho and Oregon to supply meat as the wildlife disappeared. Grant traded emigrants one sturdy ox for two worn-out beasts, then let the new animals fill out on the rich grasses in the valley. By 1862 Grant was prosperous enough to erect a spacious two-story house near his cabin, which became a bunkhouse. But Quarra kept her tepee in the front yard, perhaps because Grant had taken other Indian wives.

Grant often sold beeves to Con Kohrs, a German immigrant who owned a string of mining camp butcher shops. Born in 1835 and reared on a farm, Carston Conrad Kohrs went to sea at the age of fifteen. In time he came ashore and found work in New York City. His family immigrated, settling in Iowa, and Con became a United States citizen in 1857. Soon afterward Con "took the California fever" and went west to seek his fortune as a prospector.

Rancher Johnny Grant (left) in 1866, the year he sold his Deer Lodge property to future cattle king Conrad Kohrs (right).
— Courtesy Grant-Kohrs Ranch

In 1862 he drifted to the new gold fields of southern Montana. Experiencing little luck in the diggings, Kohrs went to work as a butcher, a trade he had learned in his early wanderings. Industrious and aggressive, he soon had his own shop, then opened others in other mining camps. His principal task was to obtain cattle for his meat markets. At a time when the region was plagued by the murderous road agents of Henry Plummer, Kohrs fearlessly rode alone carrying gold to purchase cattle. He often bought worn-out animals from travelers and fattened them on the open range for his butchers. When he located a herd somewhere, he had to learn to arrange credit. A tireless rider, he purchased a dozen of the finest mounts in the territory and stabled them at various locations along his most frequently traveled routes. In 1862 Kohrs galloped up to Johnny Grant's big house only moments ahead of pursuing highwaymen.

Grant had built Montana's first sizable cattle herd, and Kohrs regularly rode to Deer Lodge to buy livestock on credit. Since Grant never kept records, he was pleasantly surprised on one occasion when the scrupulously honest Kohrs brought him $1,100 to settle his account. In 1866 Grant agreed to an offer from Kohrs for his ranch, "farmhouses with household furniture, stables, corrals, ricks of hay, all my farming implements, wagons . . . cattle, sheep, goats and grain." The two friends agreed upon a price of $19,200; Kohrs paid $5,000, with the balance due in the spring of 1867. Grant then moved back to Canada.

Kohrs had obtained a superb base of operations for his meat market herd, although there was no title to the unsurveyed land. He filed a 160-acre homestead claim around his complex of buildings, then gradually bought or leased surrounding property totaling 33,000 acres. He rapidly acquired more cattle as well, continuing his far-flung purchasing expeditions. He branded a C on the left shoulder and a K on the left thigh, and his home spread began to be called the CK Ranch.

Now in his thirties, Kohrs determined to bring a bride to his new home. He went east in 1868 and, in Cincinnati, located nineteen-year-old Augusta Kruse, whom he had known when she was a child. Of German background, the beautiful Augusta was as hard-working as Kohrs, and, at six feet, nearly as tall (he was six-foot-three). After a whirlwind courtship she married Kohrs, then journeyed to the Montana frontier. The honeymoon trip took seven weeks by riverboat and another six days by wagon. Augusta immediately put the big house in order, then began a relentless round of cooking, cleaning, roasting coffee, milking cows, making soap and candles, and rearing the three daughters and a son she bore her husband.

In 1861 Johnny Grant erected a two-story clapboard house regarded as "the finest in Montana." The house faced east because of the Indian belief that a tepee entrance should face the rising sun.

The icehouse was built in the 1880s. Ice cut from frozen ponds was stored here for summer use. When the 1930s brought refrigeration, the structure was converted to a tack room.

Built in 1885, the Leeds Lion barn was named after a famous English Shire stallion. Around the ranch were several stallion barns; each stabled only one stallion, to prevent fighting.

A jack-legged fence, typical of the era.

Kohrs brought in his half-brother, John Bielenberg, to run the ranch at Deer Lodge, while continuing his ceaseless business travels. A lifelong bachelor, Bielenberg genuinely enjoyed working with livestock and cowboys, and he efficiently operated the home ranch for more than half a century. Within a few years registered Herefords were purchased to improve the cattle herd, and Clydesdales and other big work horses were raised and sold in impressive numbers.

Kohrs, meanwhile, greatly expanded and diversified his business interests. He invested in mining, real estate and water rights, while forming numerous partnerships with other ranchers. Kohrs thus was able to ride out market fluctuations and bad weather that wiped out less resourceful ranchers. His registered cattle, sheltered by the valley around the CK Ranch, survived the infamous winter of 1886-87 in good condition, although his herds in eastern Montana were devastated.

With vast areas of the western range suddenly vacated, and with his excellent credit and market connections, Kohrs was in position to expand. He established large grazing ranges at more than half a dozen locations in Montana, along with big pastures in Idaho, southern Wyoming, northern Colorado, and just across the border in Canada. With Bielenberg running the home ranch as a stock breeding center, Kohrs cattle grazed on more than a million acres of northern rangeland, and at the height of operations more than 1,000 cow ponies wore Kohrs' Dutch K horse brand.

Texas longhorns resting in a Grant-Kohrs corral.

During the quarter century following the winter of 1886-87, Kohrs and Bielenberg shipped at least 8,000 to 10,000 head of cattle annually. Con Kohrs, a pioneer rancher who got his start during the early days of the open range, adjusted admirably to the new era of western stockraising and became "King of Montana Cattlemen."

Because of his prominence, inevitably Kohrs became involved in public affairs. While serving on the first executive committee of the Montana Stockgrowers Association at Miles City in 1884, he met Theodore Roosevelt, whose Maltese Cross cattle ranged into eastern Montana from Dakota Territory; in later years the old friends hosted each other at the White House and Deer Lodge. Kohrs was a commissioner of Deer Lodge County, a member of the Montana Territorial Legislature, a delegate to the Montana Constitutional Convention in 1889, and a member of the Montana State Senate. Through the years many political meetings were held at the CK Ranch house, and in 1890 Kohrs added a large brick wing to his home.

Periodically, Kohrs and his family enjoyed extended vacations: to New York, Chicago (during the World's Fair), Philadelphia (during the 1876 Centennial), and other eastern cities; to New Orleans for Mardi Gras; to Yellowstone for an 1883 camping trip; and to Europe. Following an 1898 sojourn in Europe, Kohrs decided to move his family to Helena, although summers still were spent at Deer Lodge. In 1901 his only son, William, died of pneumonia while a student at Columbia University.

With the obvious heir to his empire now deceased, an aging Kohrs and John Bielenberg (ten years younger than his half-brother) continued their successful methods of operation. Indeed, cattle sales in 1909 exceeded $500,000, a record for Kohrs, and the Great War of 1914-18 sent beef prices soaring. Then, with the foresight that had distinguished his career, Kohrs anticipated the postwar market drop and sharply reduced his property holdings and herds immediately after the war.

By the time Kohrs died in 1920 in Helena at the age of eighty-five, he had sold all but 1,000 acres around the home ranch. Augusta maintained her house in Helena, but continued to spend her summers at the ranch. She contributed generously to charities and regularly visited New York City to attend the Metropolitan Opera. In 1945, at the age of ninety-six, Augusta died in Helena.

In 1932 the remnants of the home ranch fell under the management of Conrad Kohrs Warren, whose parents were Dr. O. Y. Warren and Katherine Kohrs Warren, youngest daughter of Con

There were numerous ranch vehicles, most of which advertised the Hereford cattle and Belgian horses raised by Kohrs.

The ornate dining room was paneled with golden oak.

John Bielenberg's fold-up bed in the bedroom he used for fifty-five years.

and Augusta. Warren bought the ranch from a family trust in 1940. An expert livestock breeder, he became widely known for his Hereford cattle and Belgian horses. Warren and his wife Nell carefully preserved the old ranch structures, the family furnishings, and other possessions, and the working documents of the ranch. Since 1977, when the Grant-Kohrs Ranch opened as a National Historic Site, the original buildings, corrals, and furnishings, as well as the magnificent setting, have rewarded visitors with a tangible reminder of the historic period that has captivated the world for a century.

D-S THE PIONEER CATTLE COMPANY

One of Con Kohrs' most important ranching partnerships involved noted Montana pioneer Granville Stuart.

Born in Virginia in 1834, Granville and a brother went to the California gold fields in 1852 as prospectors. Drifting into Deer Lodge Valley five years later, they discovered gold and triggered Montana's first rush. In 1860 Granville was one of several men who traded for trail-worn oxen and brought them to Deer Lodge to recuperate on the valley's lush grasses. He moved to Bannack in 1862 and to Virginia City the next year, and apparently was involved in the vigilante activities that wiped out the Plummer gang. Although active in public affairs, Granville Stuart never prospered, and by 1879 he was working as a bookkeeper in Samuel Hauser's First National Bank of Helena. Stuart had known Hauser since his arrival in Montana as a miner in 1862, and the two old friends now became excited about the possibilities of open-range ranching.

A part owner of the bank was Andrew J. Davis, who had made a fortune in mining. Davis and his brother Edwin became interested in Stuart's ranching proposals, and the four men formed Davis, Hauser and Company, capitalized in August 1879 at $150,000. The Davis brothers invested $50,000 each; Hauser borrowed $30,000 from Erwin as his share; and Stuart received a loan of $20,000 from the First National Bank for his share. Stuart would serve as general manager at an annual salary of $2,500. His first tasks were to locate a range and find cattle to bear the DHS brand, which would become one of Montana's most famous brands.

Stuart abandoned the Montana tradition — which he had

helped originate — of ranching in deep mountain valleys where there was ample water and winter shelter. Instead he sought rich grasslands on the open ranges of eastern Montana, which offered less water and shelter. Stuart decided to locate DHS headquarters at Ford Creek, a few miles from the foot of the Judith Mountains and well to the east of any previous Montana ranch. The entire $150,000 capital was spent to purchase about 7,000 cattle, which were delivered to the preempted DHS range by October 1880.

Also by October 1880, the DHS ranch buildings neared completion. Several line cabins were placed around the general range limits. At headquarters two large log cabins, one for Stuart and his family and one for a bunkhouse, were erected forty feet apart. These cabins were loopholed and connected at the rear by a log palisade, and at the front by gates so that a dozen horses could be corralled between the buildings. Stuart filled his cabin with his library of several thousand books, volumes which he shared with the cowboys who were interested in reading.

The first winter on the new ranch was severe, and Crows from a nearby reservation frequently hunted DHS cattle, just as they had the fast-disappearing buffalo. Despite these losses the DHS made its first sale of cattle in August 1881: 1,266 steers were sold to Con Kohrs for $31 per head, totaling nearly $39,000. Kohrs was expanding onto the eastern plains, and in 1883 he bought into the DHS. Kohrs formed a syndicate to purchase the two-thirds interest owned by Andrew and Edwin Davis. The DHS was valued at $400,000, and the Davis brothers were paid $266,667. The new company was named Kohrs, Stuart and Company. Kohrs owned one-third; Hauser retained an interest; and Stuart, who still owed for his share, continued as general manager.

The DHS calf crop for 1883 totaled 3,376, and the sale of 793 steers brought in over $40,000. By this time rustlers were active among the open-range herds, and a major vigilante movement apparently was led by Stuart. Another problem was the rapid overstocking of the open range. By 1884 Stuart warned a cattlemen's convention that the Montana ranges were "absolutely overcrowded." But herds continued to be driven to eastern Montana, and range conditions deteriorated rapidly. In January 1885 Kohrs and Hauser reorganized the DHS, creating the Pioneer Cattle Company and issuing 10,000 shares at $100 apiece, a paper value of

STUART'S STRANGLERS

At the Montana Stockgrowers Association meeting in Miles City on April 20, 1884, Theodore Roosevelt and the Marquis de Mores led complaints about the brazen activities of cattle and horse thieves. Cattlemen angrily called for an all-out "rustlers war," and a week and a half later a group of ranchers met with Granville Stuart at DHS headquarters. Even though Stuart's role in subsequent events was hazy, the fourteen vigilantes which organized became known as "Stuart's Stranglers."

On June 25 two rustlers stole seven horses from the string of a cowboy working for J. A. Wells. Rancher Bill Thompson gave chase, fatally wounded one thief, then captured the other man and turned him over to Wells. That night vigilantes hanged the rustler.

On July 3 word reached the DHS that rustler Sam McKenzie had been sighted nearby. The entire crew was mobilized, and McKenzie soon was brought bound to the bunkhouse. After dark he was taken out and hanged, and the next day the swaying corpse was "the center of an admiring concourse of flies."

After hearing of McKenzie's lynching, two other rustlers rode into Lewiston the next day, arrogantly spoiling for trouble. Red Owen and Rattlesnake Jake Fallon drunkenly ignited a shootout in Lewiston's only street. One citizen was killed and another wounded, but both outlaws were gunned down by an angry populace. Also on the Fourth of July, the vigilantes divided into two groups and sought out more rustlers. Red Mike and Brocky Gallagher were found just after crossing the Missouri River with a stolen band of horses. Following a ten-mile chase they were overtaken and hanged. That night the other vigilantes confronted Billy Downs and California Ed at their hideout on the Missouri. Butchered meat, a pile of fresh cowhides, and twenty-six horses with familiar brands were found at the hideout. Downs and California Ed were hanged in a group of cottonwoods.

The vigilantes then decided to assault the principal outlaw stronghold on the Missouri, fifteen miles east of the mouth of the Musselshell. There was a large log cabin, loopholed for defense,

(Continued)

(Continued)

along with an adjacent stable and corral, as well as a tent made of wagon covers and located one hundred yards away beside the riverbank. On the night of July 8 five vigilantes surrounded the cabin and three covered the tent, while the ninth man held the horses. The vigilantes opened fire at dawn, when a man emerged from the cabin. Six men scrambled out of the tent: one escaped clean; the "boss outlaw," John Stringer, was cornered and went down fighting; and four men concealed themselves in the brush, later to fashion a log raft and drift downriver. The five men inside the cabin resisted stoutly, until the building was set ablaze. As the outlaws burst out of the burning cabin they were gunned down.

The next morning the four men on the raft were captured by the military, then turned over to a deputy U.S. marshal — and a posse of DHS riders. At the mouth of the Musselshell a party of masked vigilantes took the prisoners. The thieves were hanged.

Nineteen rustlers had been killed, and stock theft in the area declined immediately. The vigilantes deliberately kept their identities obscure, but Stuart was elected next president of the Montana Stockgrowers Association. "Once I heard a woman accuse him of hanging thirty innocent men," related Stuart's son-in-law, Teddy Blue Abbott. "He raised his hat to her and said, 'Yes, madam, and by God, I done it alone.'"

$1 million. Kohrs served as president, Hauser was vice-president, and Stuart remained general manager.

The Pioneer Company continued to buy cattle, and the 1885 calf crop totaled 4,280. But these animals were grazing a ruined range, and the winter of 1886-87 took a terrible toll of weakened herds. Although the Pioneer estimated a herd size of 40,000, there probably were no more than 20,000 cattle wearing the DHS brand on the eve of winter. The roundup of 1887 produced only 7,000 Pioneers, a disastrous loss of sixty percent.

Stuart was stunned at the pitiful sight of thousands of dead cattle. Still in debt for most of his $20,000 loan, he was replaced as manager by Percy Kennett. "A business that had been fascinating to me before, suddenly became distasteful," wrote Stuart. "I never wanted to own again an animal that I could not feed or shelter."

With overgrazing no longer a problem in eastern Montana, Con Kohrs was convinced that he could rebuild profitably. Davis offered Kohrs $100,000 without security for his recovery efforts, and early in 1888 Kohrs used this capital to buy 9,000 steers in Idaho, shipping most of them to his home ranch but sending a small number to a promising range south of the Milk River near Bowdoin in northeastern Montana. "In the fall we shipped about 400 CK's and about 600 Pioneers from Bowdoin," he wrote.

Still operating in 1899, the Pioneer Cattle Company bought the old N Bar N Ranch on Prairie Elk Creek in eastern Montana (Dawson County). Kohrs supervised the remodeling of the foreman's house and the construction of a bunkhouse and "cottage." Kohrs' oldest daughter, Anna, had just married John M. Boardman, who was made general manager of the company by Con. The company held a lease on the Fort Peck Indian Reservation to the northwest and controlled nearly a million acres south of the Missouri River. For the last eighteen years that Kohrs remained in the cattle business, the Pioneer Cattle Company provided the major scene of his operations. When he was seventy-eight, in 1913, Kohrs and Anna fashioned his autobiography during a summer spent at the Dawson County headquarters.

XIT XIT IN MONTANA

John V. Farwell of Chicago, head of the Capitol Syndicate which organized the XIT Ranch in the Texas Panhandle, ordered experimental grazing on more northerly ranges. Other large Texas ranches already had learned that young steers could be fattened for market with impressive results on the grasses of Wyoming and Montana. In 1890 Farwell bought a small ranch in Custer County, Montana, on Cedar Creek sixty-five miles north of Miles City, to use as a northern headquarters. Farwell leased two million acres as a finishing range, and a subdivision was established at the old Hatchet Ranch, about twenty miles north of Terry. Word was then sent to Texas to move 10,000 two-year-old XIT steers in five herds to Montana.

O. C. Cato was sent up from Texas to manage the Montana operation. According to one of the first trail hands from Texas, Al

Veteran XIT cowboy J. K. Marsh, nicknamed "Bugger Face," sawing meat.
—Courtesy Prairie County Museum, Terry, Montana

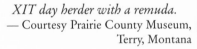

XIT day herder with a remuda.
— Courtesy Prairie County Museum, Terry, Montana

XIT cowboys around chuckwagon, with their remuda in the background.
— Courtesy Prairie County Museum, Terry, Montana

Denby, Cato was such an expert cattleman that he "could very near tell the color of a cow by looking at her track."

Another Texas drover, 250-pound Bob Fudge, hired on with the XIT after his wife died in 1894. Fudge helped drive two herds up from Texas, then stayed with the XIT in Montana for eighteen years, most of the time in charge of the Hatchet subdivision. Endorsing the policy of finishing Texas cattle on northern ranges, Fudge observed that "when they had been in Montana a year or two, they grew in size and weight to more than double the steer which we brought up the trail from Texas."

During his first winter in Montana, Fudge was the sole cowboy assigned to a lonely line camp on the Yellowstone River, the southern boundary of XIT range (the northern limit was the Missouri River). The XIT provided winter hands with fur coats, fur caps, and fur chaps. Fudge was therefore protected from the cold, but after a couple of months on his solitary rounds he was badly injured when his horse fell. As a result O. C. Cato initiated a policy of keeping two riders at each line camp throughout the long Montana winters.

These XIT horses are wearing heavy winter coats, and the XIT cowboy at right is wearing expensive angora chaps.
—Courtesy Prairie County Museum, Terry, Montana

Aside from line riding, little work was available during the winters until April, when the horse roundup began seven months of intense activity. The fall beef roundups across the vast open range of the XIT involved five big wagon crews, consisting of a chuckwagon and cook, a bed wagon carrying tents and bedrolls, eight to ten cowboys, and two horse wranglers, a day herder and a night hawk. Each roundup crew made two rail shipments during the fall, with Rufe Morris' wagon often making up three trainloads. The final shipments were dispatched by rail to Chicago by the first of November.

Although most cowboys then were laid off until the following spring, they were permitted to pick out a horse from their string to use through the winter. (Any man not allowed to keep a winter horse was wise to look for work outside the XIT.) The remaining 600 horses were kept at the headquarters ranch during the winter months. Wagon boss Rufe Morris was in charge at headquarters, keeping only a cook and one cowboy; Bob Fudge ran the Hatchet subdivision, with a cook and Al Denby.

Most of the laid-off cowboys stayed in the log bunkhouses at the two XIT ranches and, according to Al Denby, "took in all the dances in the country during the winter." Key figures at each dance were XIT men Ed and Louis Weisner, who rode to the festivities with their fiddles strapped to their backs. It was customary for every man at a dance to pitch in a dollar, and this collection was given to the musicians. Cowboys readily rode horseback twenty to sixty miles to attend a dance. If a blizzard struck, the ball might become a three- or four-day party. "But there was always plenty to eat," recalled cowboy J. K. Marsh, "as all of the married women always brought plenty of cakes."

The most popular married woman was Mrs. H. J. Kramer of Fallon, whose husband assisted the XIT with shipping. She loved to waltz and taught the dance to many eager cowboys. As one of her partners reminisced, Mrs. Kramer "was a large, strong lady who could steer them around."

"Girls were very scarce," explained Al Denby, who was one of the lucky men to find a bride. "It was very seldom that a school marm taught the second term without annexing some cowboy for a husband, and as a rule they had the entire bunch to pick from."

The primary attraction of the XIT for cowboys was the opportunity to practice the old-fashioned art of open-range cowpunching. The XIT in Texas was tightly fenced and cross-fenced, along with other ranches, large and small, throughout most of the rest of the West. But in Montana, except for horse pastures, the XIT ranges were unfenced, and young cowboys could live and work in a manner that elsewhere had largely disappeared.

At the height of Montana operations there were as many as 65,000 XIT steers on the range, along with 1,000 cows brought up from Texas to raise beef for the roundup wagons and line camps. Two long creeks, the Redwater and the Big Dry, meandered across the heart of XIT range. Although XIT steers ranged as far west as

WOLFING ON THE XIT

XIT cowboy Charlie Clement with hounds used to hunt wolves.
— Courtesy Prairie County Museum, Terry, Montana

Gray wolves, sometimes numbering as many as forty in a pack, killed upwards of several hundred XIT cattle per year. Handling cattle the same way as a group of cowboys, wolves would run the steers into a draw, then hold the bunch while the biggest lobos would go in to do the killing. The steers would use their horns to fight off attackers from the front, but a wolf would slip in behind and hamstring the prey. Hungry wolves immediately began tearing flesh from the felled victim.

"I have seen cattle and horses with great holes eaten in their hams and shoulders by the wolves," stated Bob Fudge, "and the poor animals would still be living but were unable to get to their feet."

The XIT company bought a score of dogs, including a pair of expensive English stag hounds, to track the wolves. During the winters XIT cowboy George King used the dogs to chase wolves. With a $5 bounty for each wolf pelt, XIT riders would eagerly break away from their duties to gallop after a wolf.

Wolves that had just gorged on a fresh kill could easily be chased down and shot with revolvers. But many cowboys preferred the challenge of roping and dragging a wolf to death.

"To chase and rope wolves was hard on horses and dangerous for the cowboys," testified Fudge, "but exciting — just what a cowboy liked!"

Branding on the XIT.
— Courtesy Prairie County Museum, Terry, Montana

Ladies visiting an XIT roundup wagon. J. K. Marsh is standing at left.
— Courtesy Prairie County Museum, Terry, Montana

XIT chuckwagon.
— Courtesy Prairie County Museum, Terry, Montana

the Musselshell River, during summer months most of the company's cattle grazed near the Redwater and the Big Dry. During a beef roundup enough steers usually could be gathered near these creeks to be driven to the shipping point. The XIT in Texas continued to send large numbers of steers north for double-wintering; in 1902, for example, 22,000 head were shipped to Montana.

But in 1909 homesteaders finally began to penetrate the XIT range, and sheep were introduced in large numbers. By this time the Capitol Syndicate had begun to sell off their Texas lands to farmers and ranchers, and as the cattle operation was reduced the Montana range no longer was needed. On October 8, 1909, O. C. Cato staged a farewell barbeque at the Hatchet Ranch, and soon the XIT liquidated its Montana holdings. Cato purchased the property actually owned by the company he had long served. The XIT operated less than two decades in Montana, but oldtimers from the area still fondly relate tales of the massive cattle ranch.

ᗯ W-BAR RANCH

"There never was but one Pierre Wibeaux . . ." admired his foreman, C. O. Armstrong, "and there is no other that can fill his place."

Today his "place" is represented by a statue which overlooks the range where he proved that he was one of the most progressive and successful ranchmen in the West.

Pierre Wibeaux was born in 1858 in Robaix, France. His father and grandfather owned and operated textile mills and dyeing works, and Pierre received a good education. An adventurous youth, he joined the French Dragoons when he was eighteen. After a year's service, he traveled extensively in Europe, then was sent to England to study the textile industry. There he met and fell in love with Mary Ellen Cooper. He also encountered numerous Englishmen who were excited about the Beef Bonanza of the American West.

In 1882, once again driven by a sense of adventure, Pierre determined to seek his fortune by becoming a frontier cattleman. His father was angry that Pierre was turning his back on the family business, but he gave his son $10,000 as his inheritance. Pierre journeyed to Chicago, a bustling meat-packing center, where he spent

weeks at the stockyards absorbing information about desirable livestock qualities, market conditions, and other information about the
cattle business.

In Chicago he met another Frenchman eager to cash in on the
range cattle boom, the visionary Marquis de Mores. Wibeaux accompanied the Marquis to the Badlands of western North Dakota,
where de Mores would establish a ranch and meat-packing plant
near the Northern Pacific Railroad. A short distance to the west, in
eastern Montana and western North Dakota, Wibeaux found
Beaver Valley to be an ideal site for cattle ranching. Varied timber
and deep draws offered protection from winter storms; grazing was
excellent; and in unusually dry summers, when streams stopped
running, spring-fed ponds always contained water.

In June 1883 Wibeaux formed a partnership with Gus Grisy,
also a native of Robaix. Grisy and Wibeaux devised the GW and G
Anchor W brands, then announced plans to place 10,000 head of
cattle along Beaver Creek. They bought 160 tons of native hay for
the coming winter. Property for a ranch headquarters was acquired
on Beaver Creek twelve miles north of the community of Beaver,
and a log cabin with a sod roof was built.

Wibeaux returned to France in the fall of 1883 to raise more
capital, from his family and from other French businessmen. Then
he went to Dover, England, where he married Mary Ellen "Nellie"
Cooper. Returning to Beaver Valley with a bride and investment
capital, Wibeaux bought out his partner, paying Grisy 10,000 francs
($2,000) and letting him keep 500 head of cattle. Wibeaux kept 500
cows, 475 steers, 200 yearlings, and 50 horses. He adopted a W-Bar
brand, which gave his ranch its name.

Unlike many open-range ranchers who grazed only steers for
market, Wibeaux ran mixed cattle from the beginning, and he provided hay for winter feed. Wibeaux thus avoided the terrible winter
losses which bankrupted other ranchers, and W-Bar cattle emerged
from each winter in better condition to be fattened for sale. Indeed,
in 1887 Wibeaux began raising the first alfalfa in the region. He
upgraded his cattle herd with Shorthorn bulls, most notably in
1897, when he purchased 350 in Canada. He improved his horse
herd by bringing in Kentucky stallions and Texas mares, hoping to
combine Kentucky speed with the toughness of Texas cow ponies.

In 1885 Pierre and Nellie Wibeaux moved from their log cabin

The long front porch of the W-Bar office.

Built in 1892, the ranch office in Wibeaux is now a museum.

Pierce Wibeaux's office desk.

into a frame ranch house which measured eighty-by-thirty-six feet. A low-roofed porch extended across the north side. The large rooms boasted wallpaper and carved wood, and Pierre's billiard and wine room was a special favorite of visiting ranchers and cowboys. Water for the house and lawn was piped in from a windmill tank, while the washroom featured a sink fashioned from a four-foot-long sandstone. Painted white, the Wibeaux home was inevitably called the "White House." Also in 1885 a stone barn was erected.

Shortly after moving into the White House, Nellie Wibeaux gave birth to a son, Cyril. During her first Christmas at the ranch, although still living in the log cabin, Nellie donned an evening gown and cooked a turkey, plum pudding, and mince pie. An admirable hostess in the White House, she once rode out to a district roundup to invite the crew to an open house, where she offered beer, wine, and cigars. When Pierre found it necessary to spend the winter of 1886-87 raising capital in France, Nellie presided over the ranch from the White House. When a female servant committed suicide, Nellie persevered in carrying out her duties.

Pierre was as handsome as his wife was beautiful. He fenced and boxed, and was an excellent horseman. His favorite mount was named Tic-Tac, but he also enjoyed driving a team of spirited trotters. Wibeaux industriously participated in the range activities of the W-Bar, and he traveled widely to investigate market conditions and to pursue sale or purchase possibilities. Because the White House was twelve miles from town, Pierre built an office with sleeping quarters in Wibeaux. (The village on Beaver Creek beside the Northern Pacific first was called Beaver, but in 1884 Gus Grisy, who was serving as postmaster, changed the name to Mingusville, after himself and his wife, Minnie. In 1895, however, a petition effort headed by Pierre renamed the town Wibeaux.)

Wibeaux was out of the country during the destructive winter of 1886-87, but his animals were in good condition and his losses were minimal. Wibeaux boldly capitalized on a situation that drove a majority of other ranchers out of business. He realized that the range suddenly was no longer overstocked, and that the death of so many beeves would drive prices upward. Reasoning further that cattle which had survived were especially hardy, Wibeaux aggressively purchased remnants of many devastated herds. In one season he

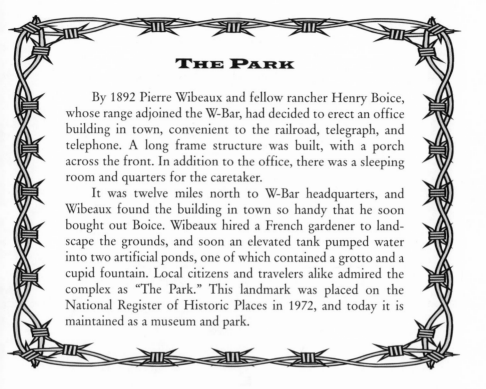

THE PARK

By 1892 Pierre Wibeaux and fellow rancher Henry Boice, whose range adjoined the W-Bar, had decided to erect an office building in town, convenient to the railroad, telegraph, and telephone. A long frame structure was built, with a porch across the front. In addition to the office, there was a sleeping room and quarters for the caretaker.

It was twelve miles north to W-Bar headquarters, and Wibeaux found the building in town so handy that he soon bought out Boice. Wibeaux hired a French gardener to landscape the grounds, and soon an elevated tank pumped water into two artificial ponds, one of which contained a grotto and a cupid fountain. Local citizens and travelers alike admired the complex as "The Park." This landmark was placed on the National Register of Historic Places in 1972, and today it is maintained as a museum and park.

made large profits from sales, while adding 16,000 head of cattle to his herd.

By 1890 his herd had expanded to at least 40,000, and for the next several years he branded 10,000 to 12,000 calves annually. It was estimated that at its peak during the 1890s the W-Bar herd numbered 65,000 head of cattle, as well as 300 saddle horses. (In 1889, when Wibeaux was offered $60 apiece for any of his horses, he sold his entire herd of 630 head, including colts, then purchased 300 remounts in a subsequent transaction.) Wibeaux usually employed twenty-five to thirty riders during the busy months, then kept ten riders through the winter. In order to combat wolves, he maintained fifty to one hundred wolf hounds, and offered a bounty of $5 and $2.50 respectively for the pelts of wolves and pups.

During the early years of the twentieth century, homesteaders entered the area in large numbers, and Wibeaux began to reduce his ranching operations. His last recorded cattle shipment was in 1905, but by then he was immersed in other business activities. In 1903 Wibeaux bought 36,000 acres of railroad land in Beaver Valley for less than a dollar per acre, then sold it the following year for $16 an acre. He built and was president of the First National Bank of Miles City; he was a major stockholder in the American Bankers Insurance Company; and he owned a mine near Deadwood. His estate exceeded half a million dollars.

But by the age of fifty-four Wibeaux had developed liver cancer, and he died in a Chicago hospital in 1913. The next year his wife and son erected the monument which Wibeaux had planned. The White House burned during the 1920s and only traces of the stone stable remain, but, like Wibeaux, his statue surveys the Beaver Valley range with a commanding presence.

In 1914 a statue of Pierre Wibeaux was erected on a hill just west of his name-sake town. The monument overlooks the old range of his W-Bar Ranch.

MONTANA
RANCHING IN A CORRAL

Biggest Ranches:
Continental Land and Cattle Company (80,000 cattle under numerous brands in eastern Montana; incorporated in St. Louis for $3,000,000)
XIT (65,000 cattle, 60 riders)
W-Bar (65,000 cattle)

Best Cowtowns:
Miles City ("Cowboy Capital of Montana") and Malta

Cowboy Legislature:
January-February 1885, Helena; passed nearly all of the statutes relating to Montana ranching.

Greatest Cowboy Artist:
Charles Russell (1864-1926); Montana cowboy during the 1880s and 1890s, later worked from log cabin studio in Great Falls.

Best Western Movie Actor:
Gary Cooper (1901-1961). Born on Montana ranch and became an expert rider while favoring a broken hip. Starred as *The Virginian* (1929) and won an Oscar as Marshal Will Kane in *High Noon* (1952).

Bloodiest Lynching of Rustlers:
During July 1884 vigilantes shot or hanged at least 16 outlaws in central Montana and recovered 85 stolen horses.

Montana Glossary:
"Westerners" . . . . . cattle from Oregon
"bannocks" . . . . . sweet fried biscuits
"chemicals" . . . . . liquor

RANCHING IN ARIZONA

Jesuit priests introduced cattle into Arizona to stock their missions and help feed the Indians. During the period of Spanish and Mexican rule, cattle-raising was one of Arizona's most important occupations.

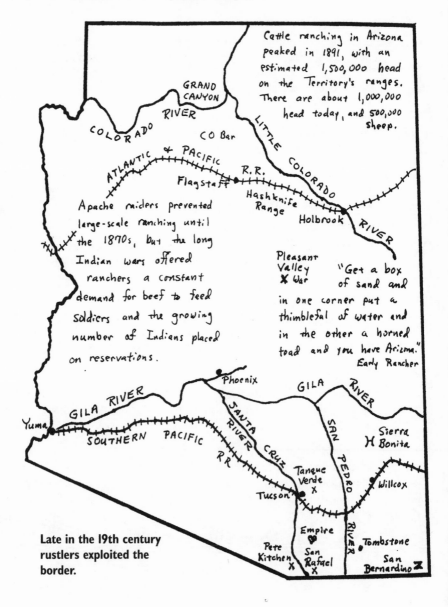

GRAND CANYON

Cattle ranching in Arizona peaked in 1891, with an estimated 1,500,000 head on the Territory's ranges. There are about 1,000,000 head today, and 500,000 sheep.

COLORADO RIVER

CO Bar

ATLANTIC & PACIFIC

LITTLE COLORADO RIVER

R.R.

Flagstaff

Hashknife Range

Holbrook

Apache raiders prevented large-scale ranching until the 1870s, but the long Indian wars offered ranchers a constant demand for beef to feed soldiers and the growing number of Indians placed on reservations.

Pleasant Valley X War

"Get a box of sand and in one corner put a thimbleful of water and in the other a horned toad and you have Arizona."
Early Rancher

GILA RIVER

Phoenix

GILA RIVER

SANTA CRUZ RIVER

SAN PEDRO

Sierra H Bonita

Yuma

SOUTHERN PACIFIC R R

Tanque Verde X

Willcox

Tucson

Empire ♥

RIVER

Tombstone

Late in the 19th century rustlers exploited the border.

Pete Kitchen X

San Rafael X

San Bernardino Z

9

Arizona

EL POTRERO RANCH

The headquarters of El Potrero Ranch was a fortress-like compound which withstood Apache raids and assaults for a decade. Pete Kitchen, the legendary "Daniel Boone of Arizona," built El Potrero in hostile territory and defended it with tenacious skill and courage.

Kitchen was born in Kentucky in 1819, later moving with his family to Tennessee. By 1846 he had drifted to Texas, where he was employed as a teamster with Zachary Taylor's Mexican War army. Moving on to California, Kitchen worked at odd jobs until he was attracted to Arizona by the Gadsden Purchase of 1854. He soon was running cattle on a spread twenty-five miles south of Tucson at Canoa Springs. In 1861, while Kitchen was delivering cattle to Fort Buchanan, a war party struck his ranch, killing all but one of his *vaqueros*, burning his buildings, and driving off 440 head of cattle and horses. For the next several years he operated a store in Magdalena.

In 1868 Kitchen returned to southern Arizona to establish a ranch just above the international boundary on Potrero Creek near its junction with the Santa Cruz River. Determined to make this new home secure for his family, he erected a virtual fortress, which he called the Stronghold, on a promontory above the creek. He

The ranch headquarters of television's High Chapparal *(filmed at Old Tuscon from 1967 through 1971) was patterned after El Potrero, including an armed lookout who was constantly vigilant atop the flat roof.*

El Potrero, the fortified stronghold of Arizona pioneer Pete Kitchen. The L-shaped adobe structure originally had a four-foot parapet atop a flat roof (this photo was taken in 1915), and two-foot-thick walls were loopholed.
— Courtesy Arizona Historical Society

built a five-room adobe house in the shape of an L, sixty feet across in front, with walls more than two feet thick. He then walled in the rest of a compound which included a well, store, smokehouse, corrals, and quarters for employees. Above the roofline of his house he constructed a four-foot parapet, equipped with loopholes and regularly patrolled by armed guards. Francisco Verdugo, who was in charge of the sentry schedule, often slept atop the roof. Laborers in the fields kept rifles slung across their plow handles.

Kitchen, Verdugo, and *caporal* (foreman) Manual Ronquillo were expert marksmen, sometimes practicing while blindfolded. Kitchen's top hand, a black man named Henry, also was a deadly Indian fighter. On one occasion, as Henry rode down a ravine, an Apache dropped a lariat onto him. With lightning reflexes Henry spurred forward and yanked the rope. The brave tumbled down and Henry pounced on him, seizing his hair and slitting his throat.

Kitchen took as his common-law wife Verdugo's sister, Rosa, who was a fine horsewoman and a crack shot. During an attack on the ranch in 1870, one sniper opened fire from behind a boulder atop a ridge 500 yards east of the house. Pete and Rosa were positioned on the flat roof behind separate loopholes, and he signaled her to fire at the sniper. Her shot narrowly missed, but the sniper jumped up to shoot before his antagonist could reload. Kitchen already had drawn a bead, however, and he squeezed off a round which killed the Apache.

Although many neighbors were slain or driven away by hostiles, Kitchen successfully fought off a succession of assaults. Pete's stepson was killed within sight of the house, and once there were three attacks within twelve hours. But refugees often sought haven at El Potrero, which became known as the only point of safety on the road from Tucson to Sonora. Apaches were especially troublesome during 1869 and 1870, but attacks and raids continued sporadically until March 1877.

In addition to raising livestock, Kitchen engaged extensively in farming. He cultivated over a thousand acres of bottomland, raising corn, potatoes, cabbage, fruits, melons, and hogs. In a typical year he sold 14,000 pounds of bacon and hams and 5,000 pounds of lard, all at thirty-five cents per pound. He made regular sales to army posts, reservations, and Tucson markets. In Tucson he bought a house, and during town visits he eagerly frequented local saloons to indulge his penchant for liquor and gambling.

In 1883 Kitchen sold El Potrero and moved permanently to Tucson. For several years he had invested in mining properties, although eventually those activities became a severe financial drain. Gambling losses also strained his finances. But he remained a colorful and popular figure until his death at the age of seventy-three in 1895. His historic old house still stands, although altered in appearance, and from 1967 through 1971 El Potrero was the model for the fortified ranch headquarters on the television series, *High Chaparral.*

E-C   TANQUE VERDE

Tanque Verde means "Green Pond" (because of the deeply colored pools beside the region's stream beds), and this well-watered area at the base of the Rincon Mountains offered a natural center for cattle ranching. The Santa Catalina Mountains loomed to the north, but lush grazing lay to the south and west. And like many other nineteenth-century cattle outfits across the West, Tanque Verde during the twentieth century has become a dude ranch, offering luxurious recreation among historic surroundings.

The first rancher to use this location was William S. Oury, a frontier adventurer who was a courier from the Alamo and a soldier at San Jacinto under Sam Houston. He fought Comanches as a Texas Ranger, survived the infamous drawing of the black beans during the Mier Expedition, and saw combat throughout the Mexican War. After the war he lived with his Mexican wife in California until 1856, when he moved his family to Tucson. Oury fought duels and killed his opponents in 1858 and in 1860, and in 1871 he was a leader of the notorious Camp Grant Massacre of approximately one hundred Arivapa Apaches. Although ranching in Arizona was extremely precarious because of Apache raids, Oury opened a small spread on the Santa Cruz River near Tucson.

In 1868, when Oury journeyed to St. Louis to bring his daughter home from school, he purchased a herd of Shorthorns, one hundred heifers and four bulls. Trailing the herd himself, he placed the animals fifteen miles east of Tucson, at Tanque Verde. The first blooded cattle in Arizona, Oury's Shorthorns contrasted markedly with the lanky Mexican longhorns and were widely admired. Apaches, however, immediately began stealing the animals, forcing Oury to hire armed guards in an effort to maintain his herd.

Despite the unrelenting Apache problem, in 1870 twenty-year-old Emilio Carrillo determined to establish a permanent cattle operation at Tanque Verde. For several years Carrillo had worked a homestead in Cochise County at Tres Alamos, where he raised corn and produced pine beams. With profits from these products he began buying property around Tanque Verde. At the foot of Tanque Verde Ridge he built a small adobe house while his fortified headquarters was under construction a short distance away. The three-room headquarters was L-shaped, with adobe walls nearly two feet thick. The floor of the family room and kitchen was of packed earth, but the bedroom boasted the area's first plank flooring, six-inch pine boards, cut in the Rincon Mountains.

The north and south wall had no openings except gunports. The east wall had two gunports and a door that opened onto a patio, where the ranch hands normally slept. The west wall, which faced the valley, was pierced by a doorway and two small windows with battle shutters. The heavy wooden doors were secured with massive iron hinges and bolts. The flat roof was surrounded by an adobe parapet. Carrillo's Tanque Verde headquarters was as stoutly fortified as Pete Kitchen's El Potrero, and Fort Lowell was only a few miles to the northwest. Indeed, the army gave Carrillo a Sharps carbine to supplement his defense efforts.

One of three nearby springs was diverted to pass through the patio and into the house. Adobe bricks were made at these springs, and an irrigated garden provided produce to eat and to sell in Tucson. Carrillo, a native of Sonora, followed Sonoran custom by planting barley atop his earthen roof, which provided additional insulation against summer heat, and he named his ranch Cebadilla, "Little Barley." Just beyond the patio were corrals, built of double mesquite posts and interwoven branches, with the capacity to pen 400 head of cattle and horses. Emilio Carrillo branded his livestock E-C, which resembled three hooks and became known throughout the Southwest as the Gancho ("Hook") brand.

Carrillo's precautions helped him to operate Cebadilla Ranch profitably. Soon he owned title to 1,000 acres of land, in addition to thirty-five sections of leased land. As his ranch prospered, he

Above: *The Tanque Verde Ranch still oper-
ates east of Tucson, but today the cowboys
rustle tourists instead of cattle. At right: The
army gave Emilio Carrillo this .50-70 Sharps
carbine to help defend his Tanque Verde
Ranch from Indian attacks. On display at the
museum of the Arizona Historical Society in
Tucson.*

*Emilio   Carrillo
erected this little
adobe in 1870 while
his headquarters was
under construction.
The old structure is
still used by the
modern guest ranch.*

*The Tanque Verde headquarters
originally was a flat-roofed,
three-room adobe fortress, with
loopholes in walls that were
(and are) nearly two feet thick.*

bought property in Tucson, eventually totaling a value of $60,000. When the danger of Indian attack ended, the gunports at the ranch house were filled in, and more windows and doors were added. The house was enlarged, and at the turn of the century an A-frame roof was built.

But danger had not passed. Outlawry remained rampant in Arizona Territory into the first decade of the twentieth century, and in 1904 bandits heard rumors that Carrillo had hidden a fortune in gold. While Carrillo was riding with one of his sons, they were overpowered and his son was tied to a tree. Carrillo would not talk, even when the bandits gradually hanged him, and they left after he fainted. There was $85,000 secreted beneath the big kitchen stove, but in saving his money Carrillo sustained a shock from which he never recovered. Emilio died in 1908 after a steady decline, and for the next twenty years Cebadilla Ranch was run by his son, Rafael.

In 1928 the outfit was sold to James Converse and Alan Gray, experienced cattlemen who changed the name of the spread to Tanque Verde. Visitors enjoyed the hunting and the handsome scenery, and like many another old working ranch, the Tanque Verde gradually was turned into a guest ranch. Today the corrals hold a large remuda of trail horses, the original adobe house stands beside a gleaming swimming pool, and the sturdy headquarters structure presides over a cluster of guest cottages.

⋊ SIERRA BONITA RANCH

"Let it rain, oh Lord. Let it rain on everything and everybody but Hooker. Cut him out. Don't give him a damn bit. Cut him out."

Although a little short on Christian principles, this prayer, or similar versions, was a favorite among small ranchers and farmers who were jealous of Arizona's most famous and successful cattle king, Henry Hooker.

A descendant of Reverend Thomas Hooker, the founder of Connecticut, Henry Clay Hooker was born in New Hampshire on January 10, 1828. In 1848 Henry found work in New York City, then moved on to Kansas City the next year as an employee of the

Sierra Bonita head-quarters. The buildings and walls were loop-holed but the house (at right) did not have exterior windows.

Main entrance to the Sierra Bonita headquarters.

Southwest corner of the courtyard. Beyond the wall at right was a large courtyard which led to the fortified stables. A windmill originally stood in the center of this courtyard.

Sierra Bonita foreman Les Shannon stands beside the enormous old bellows which now hangs in the courtyard.

Indian Department. Attracted to the California gold fields in 1852, he prospected and operated a hardware store before returning East to marry Elizabeth Hockwell in 1856. Back in Placerville, California, three children were born: Ida, Edwin, and Joseph. But about 1866 fire swept through much of Placerville, and Hooker lost his store.

He saved about $1,000 in cash, however, and boldly bought several hundred turkeys at $1.50 apiece. With a hired hand and several dogs, Hooker trailed his turkey herd across the Sierra Nevada Mountains toward Nevada's booming Comstock Lode, where he hoped that miners would pay for a change in diet. There was an anxious moment when the dogs ran the turkeys over a precipice, but the birds fluttered safely to the bottom. In Carson City Hooker found a market, selling out for $5 a bird, and with his turkey profits he would build a notable cattle empire.

Hooker invested these profits in longhorns, which he bought cheaply in Texas and New Mexico, then sold at a good price at Camp Goodwin in southeastern Arizona. Beef contracts were readily available for Arizona army posts and Indian reservations, because there was little ranching as a result of constant Apache depredations. Hooker embraced this opportunity, accepting government contracts for beef and other supplies. He traveled from post to post to compile a list of needed supplies, protected by a Mexican guard; Hooker drove a buckboard on these lonely journeys, and he had the back seat reversed so that his guard could face the rear.

On one of his trips Hooker encountered Cochise, who claimed that he could have killed him many times, but Hooker brought many cattle to the region — for Apache braves to steal. Hooker prudently made it a practice to give beef cattle to any nearby Apaches, and he presented Cochise a red blanket with his name woven in it. Cochise had a large red blanket woven in Mexico for Hooker, decorated with the initials "HCH" and the cattleman's Crooked H brand. After the death of Cochise in 1874, the great chief reputedly was buried in the blanket given him by Hooker.

In 1872 Hooker purchased 15,000 longhorns in Texas and Mexico to fill his contracts. With a few men he held a remnant in the Sulphur Spring Valley, but the half-wild animals stampeded one night. The animals were found in a grassy meadow in the northern

GUESS WHO'S COMING
TO DINNER?

Henry Hooker's hospitality became legendary in Arizona. He enjoyed entertaining, and the Sierra Bonita offered an excellent table, good hunting and fishing, hot mineral springs, and horse racing. Hooker insisted that men wear coats and ties at dinner, and since such finery was not universal on the Arizona frontier, he kept a supply of these haberdashery items on hand. He also bought dozens of the warmest and softest beaver and buffalo robes, so that his family and friends could bundle up on long winter rides.

Initially, Hooker hosted army officers and their wives from nearby Fort Grant. The military provided a major portion of Hooker's business, and the Sierra Bonita gave the officers and their wives a welcome break from the monotony of garrison life. Generals George Crook, William T. Sherman, Nelson Miles, O. O. Howard, and William Shafter enjoyed the comforts of the Sierra Bonita, along with a host of lesser officers.

Famed naturalist John Muir was a guest, and artist Frederic Remington sketched Sierra Bonita cowboys and livestock during an extended visit. At Sierra Bonita and Fort Grant dramatist Augustus Thomas found color and people and incidents with which to create his play, *Arizona.* He modeled a character upon Hooker, and upon Hooker's daughter-in-law, as well as the Chinese cook and a Sierra Bonita *vaquero.* The immensely popular play spread the fame of the ranch, and the Sierra Bonita became a tourist attraction.

end of the long valley. In 1868 and again the next year, Hooker had been forced out of attempted ranching locations by Indians. But when he saw the lush grasses and abundant water of the northern Sulphur Spring Valley, he determined to establish a permanent ranch. He promptly filed a 160-acre homestead claim on the area where he would build a fortified headquarters.

Hooker erected an adobe house eighty feet square around a patio. The exterior walls were sixteen feet tall and twenty inches thick. The flat roof inclined slightly toward the patio for runoff, and interior rooms opened off the patio on three sides. The fourth side was a wall which opened onto a walled courtyard, with an adobe stable opposite the patio wall. A heavy gate gave access to the courtyard, but there were no doors or windows in the exterior walls of the house — only gunports. The exterior walls were three feet taller than the roof, providing a parapet for riflemen. A windmill in the patio furnished a reliable water supply (another windmill and a 20,000-gallon tank would be added near the house), and a storeroom and root cellar were amply stocked with food. During the next several years a number of Hooker's cowboys were killed by Apaches on the range, but his fortified *hacienda* proved a safehaven from Indian dangers.

Hooker built an adobe bunkhouse near his *hacienda*, while a large wooden barn housed part of the crew in a room beneath the gabled roof. Big red letters on the barn proclaimed "SIERRA BONITA RANCH 1872." There was a large main corral 300 feet square, in addition to a smaller corral and an adjoining enclosure. Near the stables 250 tons of hay were stacked in another corral. The corrals were built of adobe bricks made from a nearby Indian burial mound; arrowheads and pottery fragments could be seen in the bricks. The stable off the courtyard boasted an innovative skylight, as well as seven stalls for stallions and space for seventeen more horses.

Hooker soon controlled a range that measured thirty miles north to south and twenty-seven miles east to west, with his Crooked H herd numbering as many as 20,000 head. At an altitude of more than 4,000 feet and with a mild climate, the Sierra Bonita ranges proved ideal for breeding as well as grazing. And when the Southern Pacific Railroad reached Willcox in 1880, Hooker had a shipping point only twenty miles south of the Sierra Bonita. He

Above the barn west of the house was a large room which served as a bunkhouse for Sierra Bonita cowboys.

A hitching post south of the Sierra Bonita barn.

This adobe stable wall was protected by loopholes.

introduced Shorthorn bulls to his herd, but later determined that Herefords gained more weight, fared better during dry spells, and could cover twelve miles while the Durhams traveled eight. Hooker paid $30,000 for 200 Hereford bull calves from Illinois, and later purchases included sixty Hereford bulls from Iowa and seventy-five from Missouri. He also placed blooded cows on his range, and eventually ninety percent of his calves were whitefaces. In order to keep scrub stock from other outfits away from his Herefords, Hooker built twenty miles of fence at either end of "his" range — most of which was preempted.

Hooker also established a dairy herd, to supply his dining room and bunkhouse tables with milk and butter. The excess was fed to his drove of Poland China hogs. And Hooker took special interest in breeding and training fast horses, roadsters as well as saddle mounts. He kept half a dozen fine stallions, and as many as 500 brood mares. Hooker built a half-mile track, maintained several paddocks, and hired trainers and jockeys. He reveled in driving to Tucson or Tombstone to display his trotters at breakneck speeds. Fined three times for driving unsafely in Tombstone, he paid up without complaint. "What I came for was to show off my horses," Hooker shrugged, "and I have succeeded in doing it."

Hooker employed a crew of more than forty men. In 1876 teenaged Henry Antrim hired on at the Sierra Bonita, but Hooker's first foreman, William Whelan, soon discharged the boy. Henry began loafing around the hog ranch at Fort Grant, engaging in petty thievery and rustling. He killed a man in a saloon fight on August 17, 1877, escaped custody, and began winning an undying reputation as Billy the Kid. "We take a man here and ask no questions," commented Hooker. "We know when he throws his saddle on his horse whether he understands his business or not . . . . A good many of our most useful men have made their mistakes."

Hooker was a friend of Wyatt Earp, who stopped by the Sierra Bonita with Doc Holliday, Warren Earp, and other gunmen involved in the bloody aftermath of the OK Corral shootout. Hooker provided the Earp party food and rest on March 27, 1882, and the next day refused to give directions to a pursuing posse led by John Behan and Ike Clanton. Years later Hooker hired Warren Earp, Wyatt's youngest brother. But Warren had developed a drink-

HOOKER'S OUTLYING RANCHES

At the height of his operations, Hooker established six outlying ranches around the Sierra Bonita home ranch, each provided with suitable buildings and corrals:

Sonora Ranch — six miles east of Sierra Bonita headquarters; irrigated by a mountain stream and used primarily for agriculture.

Cinega Ranch — four miles north of Sierra Bonita; partially used for grazing, but a significant portion was under cultivation, in part as a vegetable garden.

Mud Springs Ranch — seven miles southwest of Sierra Bonita; provided water and grazing for 5,000 cattle.

Riley Ranch — a mile southwest of Mud Springs Ranch, on an open mesa (once used as a stage station).

High Creek Ranch — located at the foot of the Galiuro Mountains, west of Sierra Bonita; used as a breeding ranch.

Box Springs Ranch — six miles west of Sierra Bonita; the springs, along with a lovely stream and handsome timber, provided a delightful summer resort.

ing problem, and in a Willcox saloon on February 6, 1900, he was shot to death by Hooker foreman John Boyett.

In 1897 Hooker decided to concentrate his operations on pure-bred livestock raised on the home ranch. He sold the herds and property on his outlying ranches, then built a fine home in Los Angeles, where he had raced his horses for years. Hooker commuted between Californa and his Arizona ranch until he died in Los Angeles on December 5, 1907, only a few weeks before his eightieth birthday.

The Sierra Bonita, now comprising just 5,000 acres, passed to his three children. Ida and Edwin bought Joseph's interest, then Ed Hooker became manager in 1910. Ed regularly purchased the homesteads of farmers discouraged by drought, and leased any available lands. Finally, 46,000 acres of deeded and leased land was under fence, with a beef herd of approximately 2,500 purebred Herefords. When Ed Hooker died in 1932 his son, Henry, took over, and today the famous Sierra Bonita still is owned and operated by Hooker descendants.

 ## THE EMPIRE RANCH

The Empire Ranch, prophetically named after the nearby Empire Mountains, was a pioneer spread that was built by Walter L. Vail into one of Arizona's largest cattle outfits.

During the 1870s, William Wakefield acquired the original 160-acre homestead, located fifty-two miles southeast of Tucson and overlooking a spring-fed stream which ran through a depression called Empire Gulch. A large adobe house was built, with four rooms and a central hallway forty feet long and eighteen feet wide. The house faced north, and adjoining the rear was a corral one hundred feet square, with a stable in the southwest corner.

Walter Vail was born in 1852 in Nova Scotia, but grew up on his family's New Jersey farm. Determined to seek opportunity in the West, he journeyed to Virginia City, Nevada, in 1875. Soon he found employment as a timekeeper in a California mine, but before the year was out he sought the advice of his prosperous uncle, Nathan Vail of Los Angeles, regarding a change of occupation. Nathan urged his nephew to enter cattle ranching, and he introduced Walter to a man named McCarthy, who wanted to put money into a ranch. But after a grueling journey to Tucson, McCarthy refused to leave town for fear of being killed by Apaches. Vail, however, actively inspected prospective ranches, and was especially impressed with William Wakefield's 160 acres in the midst of an inviting range.

Vail returned to Virginia City hoping to earn enough money to buy the property, then again sought help from Uncle Nathan, who introduced him to wealthy Englishman Herbert R. Hislop. Hislop had come west seeking a healthful climate and an opportunity to invest in the booming cattle industry, and in the summer of 1876 he headed for Tucson with Walter Vail.

During Vail's absence from Arizona, William Wakefield had sold his land and cattle for $500 to his brother-in-law, Tucson merchant and land promoter E.N. Fish, and Fish's business partner, Simon Silverberg. Fish offered the Empire Ranch for $3,800, but accepted $1,174 from Vail and Hislop. Vail asked his uncle for a loan to cover his half of the investment, and the sale was made on August 22, 1876. Hislop wrote enthusiastically that "everybody says it is the best ranch in the territory and I believe they are right."

The purchase of Vail and Hislop included 612 cattle and a small horse herd. Within a few weeks, recognizing the need to expand, the partners bought an adjoining sheep ranch, paying a dollar a head for 620 sheep and $554 for 160 acres of rich bottomland. They cared nothing for the sheep, but hired a Mexican herder to tend their flock. Looking around for other property, Vail and Hislop realized that continued expansion would require more capital, and they extended their partnership to John N. Harvey, another Englishman sent to Tucson by Nathan Vail. The firm changed its name to Vail, Hislop, and Harvey, but neighboring settlers dubbed the Empire the "English Boys' Outfit." The horse brand was a "VH," while cattle were worked with a heart brand.

MERRY CHRISTMAS

Herbert Hislop's first Christmas on the ranch was 1876, and he prepared a traditional holiday feast for the men on the new spread. "When work was over Christmas day we set down to a festive meal," he wrote. There was no turkey, but four wild ducks stuffed with bread crumbs provided a delicious main course. And the pièce de résistance was a plum pudding decorated with a sign, in English and in Spanish (for the Mexican sheepherder), which read, "Wishing you all a Merry Christmas." Hislop surrounded the pudding with brandy and lit it.

Horses were grazed in a large, fenced pasture near the house, and were penned at night in the corral behind the house. For a decade, until Geronimo's final surrender in 1886, the Empire horse herd was vulnerable to Apache theft. "When one gets up in the morning he does not know whether he will be killed by these murdering, plundering Apache Indians or Mexicans," wrote Hislop.

There were no doors or windows in the openings of the unfurnished adobe ranch house, which had a dirt floor and a leaky dirt roof. "None of the ranchmen seem to possess beds," observed Hislop, "everybody lies on the floor." The summer heat encouraged sleeping out of doors, where "the accommodation is quite as good out as in. . . ." In the absence of coal, mesquite was used for fires, "but it is abominably hard to cut." After killing a rattler under the stove, Hislop related: "In our bedroom we have lizards and bats, but we do not mind these. . . . The things I object to most are snakes but they will soon lay up for the winter." As soon as affordable, doors and windows were installed, the interior walls were plastered, and rudimentary furniture was acquired. Hislop even paid $10 for a large tin bathtub in Tucson, and he took a cold bath every morning.

In January 1877 the partners bought an additional 793 head of cattle, disposing of their sheep in the transaction. Later in the year Vail traveled to John Chisum's ranch in eastern New Mexico, bought forty Durham bulls, then drove the animals through Apache country to the Empire, aided by Chisum cowboys. Although Hislop wanted to market some of their livestock as a partial return on their investment, Vail and Harvey insisted on delaying sales until their herd had produced several calf crops. Tired of endless work in a dangerous and primitive environment, Hislop decided to return to England. Vail borrowed $6,850 from his Aunt Anna, Nathan's wife, and bought Hislop's interest in the ranch. When Hislop left he vowed never to return to "this bloody country again."

Later in the year Vail won election as the delegate from Pima County to the Arizona Territorial Legislature, becoming at twenty-seven the youngest member of the Assembly. During his single term, Vail introduced a measure which repealed a Pima County fencing ordinance, replacing it with a measure requiring all cultivated land in the county to be fenced with expensive stone or rail walls. Agriculture thus was discouraged, since farmers could not afford such expensive fencing. Vail's legislation had turned a restrictive fencing law in favor of cattlemen.

TOTAL WRECK

Walter Vail seized an opportunity to diversify his holdings in 1881 by developing the Total Wreck silver mine in the Empire Mountains. With his uncle, Nathan Vail, Walter bought the Total Wreck mining claims for back taxes.

When John L. Dillon discovered the silver-lead mine in 1879, he was asked what he intended to name his claim. Dillon commented that the boulder-strewn hill where he had found paydirt looked like a total wreck, and the name stuck. But Dillon accomplished little development, giving Vail a silver-lined opportunity near his ranch headquarters.

Nathan Vail came from California to help manage the mining effort. A seventy-ton mill was built, and by 1883 the booming town of Total Wreck boasted a population of 200. There were fifty houses in Total Wreck, along with three hotels, three stores, a butcher shop, a lumberyard, and four saloons.

The first inhabitants of the Total Wreck cemetery were six Mexican woodcutters, who were jumped by Geronimo's warriors in June 1883 while they were cutting fuel for the mill on the west side of the Whetstone Mountains. Narrowly escaping the graveyard was E.B. Salsig, who was shot in the chest during an altercation in Total Wreck. But the bullet lodged in a large packet of love letters. Fittingly, Salsig married the sweetheart whose correspondence saved his life.

The Total Wreck produced $500,000 worth of bullion before a lengthy silver depression began in 1884. Production rapidly declined, the mill shut down, and the post office closed in 1890, but Walter Vail and his uncle had earned an enormous amount of capital to use in expanding the Empire Ranch.

By 1879 the booming mining district offered a ready demand for beef, and the Empire began making its first major sales. Miners would buy all of the beef the Empire could raise, regardless of quality, and Vail commenced a brisk trade in inferior Mexican cattle, mostly three- and four-year-olds. By the end of 1879 the Empire herd exceeded 2,200 head, and Vail was able to make payments on his notes.

The Southern Pacific Railroad built through Arizona north of the Empire in 1880, further stimulating the cattle business. The Empire shipped out of Pantamo on the Southern Pacific line. Vail more than doubled the size of his herd during 1880, and during the next few years he bought a number of adjoining ranches. These property purchases created a web of partnerships with several ranchers. Nathan Vail also was a reliable source of capital for his nephew, and in

This 1896 Visalia stock saddle was owned by Harry L. Heffner, foreman of the Empire. It was trimmed with silver conchos from Mexico pesos.

1881 the two men began developing the Total Wreck mining claims near the ranch, realizing considerable profits from the venture.

By this time Walter Vail had acquired the skills of a cowboy. "He could ride anything," recalled Harry Heffner, a turn-of-the-century Empire foreman. "He was a good horseman. Good cowman. He was a man that never told you to go out to do something. He'd say, 'Come on. Let's do it.'"

Vail spent long hours in the saddle tending his cattle, but as soon as he arrived on the Empire he hired two ranch hands. His crew rapidly increased, made up mostly of Mexican *vaqueros*. Vail often hired men suspected of rustling, on the theory that they would not steal from him. He also hired area homesteaders, then would buy their property. Vail's first foreman was rugged Tom Turner, a no-nonsense Texan. Hard-working Empire cowboys claimed that they only had a chance to go into town about once a year, and all were single. Vail didn't give them time to get married.

But Vail took time to get married in 1881, to Margaret Newhall, a propertied woman from New Jersey who purchased John Harvey's interest in the Empire. Vail already had added a line of adobe rooms to the rear of his original house, and now for his

The original head-quarters of the Empire Ranch was a four-room adobe. Originally the windows were empty openings, uncovered by glass, shutters, or curtains.

The residential addition to the Empire headquarters included a bay window — a notable feature when glass was still uncommon in frontier Arizona.

Outbuildings south of the ranch house.

RIDIN' AND ROPIN' ON THE EMPIRE

"This country is famed for bucking horses and no others can buck like them," declared Herbert Hislop shortly after moving onto the Empire Ranch. "It is a funny sensation, I did not know whenever I was going up in the air. . . . I am not anxious to get on [a bucking horse] again, as I am not quite tired of my life."

Hislop bought a mare named Madge for $40 and rode her seventy miles in two days with only one feed of grass; however, she showed no signs of fatigue. The favorite mount of Walter Vail was Sheep, a bay. Another big bay, Warrior, was the most famous horse on the ranch. A superb cutting horse, Warrior belonged to foreman Tom Turner. During roundups rope artists Tom Wills or Blas Lopez would lasso whatever horse a cowboy called from his string.

Cowboys on the Empire used riatas, not grass ropes. They made their own, about forty-five to fifty feet in length, although Tom Wills used a sixty-five-foot riata (and some admirers swore it was a hundred feet long). Empire cowboys lassoed calves by the neck, not the heels.

Each man had eight to twelve horses in a string, but they changed strings often to work another group of mounts. Empire cowboys were expected to shoe their own horses. Everybody used spade bits, and army blankets, not Najavos, served as saddle blankets. Tucson's leading saddlemaker, a craftsman named Villascuso, made all of the saddles used on the Empire. Normally the ranch kept about 300 saddle horses, but many wild horses roamed the vast Empire ranges. A sweeping cleanup produced 4,000 wild horses, sold to a buyer at Pantano for a couple of dollars a head.

bride he built an expansive residential addition further to the rear. Within a few years the original four-room adobe had become a sprawling collection of twenty connected rooms.

By 1882, 10,000 cattle grazed on Vail's ranges. Cattle and property expansion had put him $40,000 in debt, but annual sales grossed $25,000 and the Total Wreck was beginning to pay. During the next few years, while he continued to enlarge the Empire, Vail helped organize the Arizona Stock Growers Association and served a term as president. Soon he became concerned about overgrazing on Arizona ranges, and he began to acquire property in California. In 1888 he entered a partnership with Carroll Gates, a Los Angeles real estate dealer with formidable business skills.

Unhappy with Southern Pacific shipping rates, in 1890 Vail sent his brother Edward, Tom Turner, and eight *vaqueros* on an old-fashioned trail drive to leased property in California. Only thirty out of 917 steers were lost during the seventy-one-day drive. Vail saved $4,000 on the expedition and encouraged other ranchers to make similar drives. Because of Vail's strategy, the railroad lowered shipping rates to previous levels.

During the 1890s, Empire ranges extended over more than one million acres, grazing an estimated 30,000 cattle. But Vail and Gates acquired extensive holdings in California and Kansas, and in 1896 Vail moved to Los Angeles, where the business offices were centered. Gates found it necessary to move to Kansas City for a few years, but when he returned to Los Angeles, the partners began to invest cattle company funds in the city's land boom. The interests of Vail and Gates became increasingly diversified — and profitable.

In 1906, while descending from a Los Angeles streetcar, Vail was struck by an oncoming streetcar. Crushed between two streetcars, the pioneer rancher who had braved Apaches and rustlers, who once had survived the bite of a gila monster, died at the age of fifty-four.

As sole surviving partner, Carroll Gates faced an overwhelming workload. Vail and Gates already had begun reducing their cattle operation, and as part of an amicable but complex estate settlement, Vail's wife and seven children assumed control of the Empire Ranch in 1909. The heirs continued to operate the Empire as a breeding ranch for two decades, maintaining a herd of several thousand Herefords. Sales were excellent during World War I, but depressed prices after the war were compounded by drought in Arizona. In 1925, with railroad rates prohibitive under existing conditions, the Empire trailed 7,000 steers down the Santa Cruz River to the Mascarenas Ranches in northern Sonora. It was fitting that the last major cattle drive in Arizona history was executed by the Empire.

In 1928 the Chiricahua Cattle Company was ordered to remove 20,000 cattle from the leased rangeland on the San Carlos

Indian Reservation. Rather than liquidate the herd, manager Frank Boice bought the Empire and other southern Arizona ranches from the Vails. The company sold a portion of the Empire Ranch in 1919. Later Mr. and Mrs. Frank Boice and their sons bought the remainder of the Empire Ranch from other company partners. The Boice family operated the Empire until 1960, when it was sold to the Gulf America Corporation for a proposed real estate development which never materialized.

A mining company bought the Empire Ranch in 1974 for mineral potential and water rights, but the company changed its plans and put the Empire back on the market. A groundswell of public support developed to preserve the historic old ranch and its natural resources in its nearly pristine condition. A series of land exchanges placed the ranch into public ownership in 1988 under the administration of the Bureau of Land Management, and today a drive down a lonely road brings visitors through Empire range to the adobe structure first inhabited by young Walter Vail during the nation's centennial.

RANCHO SAN RAFAEL DE LA ZANJA

Originating in 1822 as a Spanish land grant of four square leagues, El Rancho San Rafael de la Zanja was abandoned by the mid-1800s because of marauding Apaches, who halted almost all ranching in Arizona. But in 1883 the old *rancho* was acquired by Colin Cameron, who ruthlessly established an impressive cattle operation, generating deep resentment by his roughshod methods. And after nearly a quarter of a century he sold the San Rafael to Bill Greene, one of the most dynamic figures in Arizona history.

There were other San Rafael land grants; San Rafael de la Zanja was located along the ditch-like Santa Cruz River (*"zanja"* means "ditch" or "trench"), which flowed south into Sonora. The 1822 survey of the grant measured three square leagues running north to south and an adjoining square league to the west, a total of 17,324 acres. But a 1791 Spanish directive for land grant might be interpreted "four square leagues" — or "four leagues square," sixteen leagues totaling approximately 70,000 acres.

After the Civil War a few ranches operated out of fortified strongholds, but cattle ranching in Arizona remained limited until 1880, when the Southern Pacific Railroad entered the territory. Introductory freight rates were based on head count rather than weight, encouraging ranchers to ship scrub cattle of any age. Rollin Rice Richardson, a Pennsylvania oil man, invested $40,000 to buy an option on the San Rafael from the Romero family (successors to the original grantee) and to purchase the range cattle of nearby squatters. But in 1883 Richardson sold out for $150,000 to the San Rafael Cattle Company, a syndicate headed by Gen. Simon Cameron and Senator James Donald Cameron of Pennsylvania.

General Cameron had served as secretary of war under President Lincoln, as U.S. minister to Russia, as U.S. senator for eighteen years, and was president of two regional railroads. His son, "Don" Cameron, was secretary of war under President Grant before stepping into his father's Senate seat. Colin Cameron, a cousin born into this prominent family in 1849, managed the Elizabeth Stock Farms in Lancaster County, Pennsylvania. An experienced cattle raiser, confident to the point of arrogance, the hard-driving Colin was an obvious choice to manage the family's Arizona ranching venture.

The powerful Camerons provided Colin with considerable support in Arizona Territory. Colin's brother, Brewster Cameron, moved to Tucson, working as attorney for the San Rafael Cattle Company and serving in various influential federal appointments, including chief clerk for Arizona's Supreme Court. Various family friends were appointed to offices from U.S. marshal (Thomas Tidball) to chief justice of the Supreme Court (Richard E. Sloan) to governor (Lewis Wolfley). And after a relative, Louis Cameron Hughes, was appointed territorial governor in 1893, Colin and Brewster "exerted considerable influence upon him." Lesser offices were not neglected; a postal clerk, for example, dutifully intercepted a complaint from a homesteader addressed to the U.S. Land Commission.

Senator Don Cameron tried to have the San Rafael Land Grant confirmed as sixteen square leagues, but Congress eventually upheld only the original four square leagues. Colin Cameron, however, aggressively expanded the company's range, often by highhanded methods. During the years of legal efforts to gain sixteen square leagues, Colin ran San Rafael cattle on this land and other nearby ranges. He presumptuously wrote a prospective investor in 1884: "We have a tract of country thirty or thirty-five miles wide by fifty long — it is a principality." Colin even tried, unsuccessfully, to claim part of the Fort Huachuca military reservation.

Squatters and legal homesteaders alike were threatened, and Cameron often filed complaints which cost them fines or at least legal fees. Friendly law officers sometimes rode with San Rafael cowboys to intimidate small operators. Reportedly, a nester named Rafferty and a Mr. and Mrs. Fritch were killed and their homes were burned. An eighteen-year-old boy driving a few San Rafael cattle was summarily hanged, and other rustlers were vigorously pursued. Colin did not hesitate to fence in public lands, sometimes blocking roadways. He commandeered range by similar methods across the border, finally prompting the governor of Sonora to issue a warrant for his arrest and to call for extradition from the governor of Arizona. In 1892 Cameron was hanged in effigy in the border town of Nogales, and an 1898 story in the *Arizona Star* was headlined: "Colin Cameron's Cruelty."

When Cameron moved to Arizona in 1883, he centered the San Rafael Cattle Company just above the border near La Noria, calling his headquarters "Lochiel" after the family's ancestral home at Loch Isle in Scotland. There was an adobe house which he expanded to seven rooms, while digging two wells and erecting corrals and outbuildings. Later he bought a house recently constructed a few miles to the northwest by mining men, enlarged it to ten rooms, and moved into it with his wife, Alice, and their son and three daughters. The family also enjoyed a town house in Tucson. The big barn housed a blacksmith shop and a machine shop, and there were five other outbuildings. Near the house were rose gardens and orchards of fruit trees, and beside the Santa Cruz River three Chinese employees produced vegetables and melons at the "China Garden."

In 1892 Cameron built a two-story adobe, with eleven rooms and four baths, and a broad porch on three sides. But in the spring of 1899 a "Settlers Protective Association" tore down several sections of Cameron fencing, then threatened to burn his corral and buildings. On the following Christmas Eve the big ranch house was burned, presumably by resentful neighbors of Cameron. The family moved back to the ten-room house, then Cameron defiantly began planning construction of a grand three-story home.

Completed in 1900, the magnificent dwelling was built of bricks fired in kilns on the ranch. The basement level had a cement floor and plastered walls, and included storerooms and a big water reservoir in the middle of the long hall. A porch flanked the main level on all four sides. The main floor contained the ranch office, a parlor, dining room, study, kitchen, five bedrooms, and three baths. The top floor, beneath the dormered roof, held four bedrooms and a bath.

Immediately upon his arrival in Arizona, Cameron introduced Shorthorns and Herefords from the East. He regularly sold off his range cattle and concentrated on building a Hereford herd. He served a term as president of the American Hereford Breeders' Association, helped form the Arizona Cattle Growers Association in 1903, was appointed chairman of the Arizona Cattle Sanitary Board, and was a prominent member of the National Live Stock Association. Cameron finished his steers on rented pastures in Oklahoma and Wyoming, and when the Southern Pacific raised shipping rates, he sent a herd to market with an old-fashioned trail drive.

Cameron was an unloved but highly successful cattle baron for two decades. In his mid-fifties, however, his health began to fail, and he sold the San Rafael in 1903. Although Cameron moved to Tucson, he continued to raise cattle in Sonora until he died in 1911, at the age of sixty-one.

The San Rafael was purchased as the crown jewel of a ranching empire built by the Southwest's most flamboyant promoter. "Colonel" William C. Greene (the title was nonmilitary, assumed by the promoter to enhance his image) had based his fortune and business enterprises on the copper mining district of La Cananea, thirty miles below the Arizona-Sonora border. A larger-than-life individual, Greene was born in Wisconsin in 1853 and went west to seek his fortune when he was nineteen. He was a surveyor, teamster, farmer, and cowboy in North Dakota, Montana, Kansas, Colorado, and Texas. By 1877 he had drifted to Arizona, where he prospected relentlessly for mineral wealth, fought Indians on several occasions, and gambled inveterately at poker, faro, and roulette.

Greene married a prosperous widow and invested her inheritance in cattle. When one of his daughters drowned, he shot to death a neighbor he regarded as indirectly responsible. While investigating old gold and silver mines in Sonora's Cananea Mountains, Greene discovered quality copper ore that he was able to develop with a series of dazzling promotions. Attracting well-heeled Eastern investors, Greene organized the Cananea Consolidated Copper Company, with himself as president and general manager. In 1902 he spent $50,000 constructing a thirty-four-room mansion in Cananea.

William C. Greene, flamboyant copper king who bought the San Rafael in 1903 for $1.5 million.
— Courtesy Arizona Historical Society

Greene's prosperity stimulated a spectacular expansion of his ranching activities. In 1901 he organized the Cananea Cattle Company, with the initial goal of providing meat for the community of Cananea, and the Greene Cattle Company, to control his Arizona range. Greene soon assembled seven Sonoran ranches of about 100,000 acres each: the Cananea, Nogales, Martinez, Turkey Track, San Fernando, San Lazaro, and Cuitaca. Just north of the border and east of the San Rafael was Greene's Palominas Division. And in 1903 he paid $1.5 million for the San Rafael, by now famous as a ranching showplace, and especially renowned for the blooded Hereford herd built by Colin Cameron.

Greene now owned or leased 800,000 acres, and in 1904 his men branded 35,000 calves. Charlie Wiswall managed Greene's cattle operations. Tough but honest, Wiswall was called "The Big Noise" by American cowboys, and "Jesu Cristo" by the *vaqueros.* But Greene took a keen personal interest in his ranches, particularly the San Rafael, with its magnificent herd of Herefords. He

Greene Cattle Company cowboys riding in to a company store.

brought his San Rafael bulls across the line to upgrade his Mexican cattle. A passionate lover of fine horses, Greene bred a stable of thoroughbreds on the San Rafael, as well as a herd of Shetland ponies, which he often presented as gifts to friends or business associates. Tom Heady was installed as manager of the San Rafael. Since Greene continued to live in Cananea, the San Rafael cowboys bunked in the Cameron mansion.

In 1908 Greene's mining empire collapsed, but he still had his ranches, even though it became necessary for him to encumber these properties with debt. When he died as the result of a buggy accident in 1911, Wiswall continued to manage the ranches for Greene's heirs, while Tom Heady ran the San Rafael for thirty-eight years, earning a reputation as one of Arizona's top Hereford breeders and carrying on Colin Cameron's tradition of fine Herefords. And after nearly a century, Cameron's mansion continues to be a reminder of one of Arizona's most historic ranches, located many miles from any paved road and so picturesque that it is sometimes used for western movies.

⌐ HASH KNIFE

Arizona's legendary Hash Knife descended from the largest cattle outfit of the Old West. For more than a decade before helping found a renowned ranch in Arizona, the Hash Knife had maintained impressive operations in Texas, as well as in Montana and Dakota territories. Hash Knife management was resourceful and progressive, but on every company ranch there were cowboys who were troublemakers and/or who threw a wide loop. And when the Hash Knife moved onto a vast range in northern Arizona, rowdiness, violence, and dishonesty continued to characterize the crew.

John N. Simpson originated the Hash Knife brand as early as 1872. Born in 1845 in Tennessee, young Simpson enlisted in the Confederate Cavalry during the first year of the war. In 1866 he moved to Texas, settled in Weatherford, and entered the cattle business. By 1874 he had decided to run cattle on open range 130 miles to the west in what would become the northeastern portion of Taylor County. The brand for this new operation would represent a hash knife, a kitchen tool used by chuckwagon cooks for chopping meat and potatoes to make hash. (Simpson probably had used this brand since 1872.)

"Simpson's *rancho*" was run from a dugout seventeen miles north of Buffalo Gap, on the future site of Abilene Christian University. There was a fight with Comanche warriors the first year that Simpson occupied this range, and rustlers were active in an unorganized province with no law. Simpson and Jim Loving, son of pioneer cattleman Oliver Loving, organized a two-day meeting of ranchers, held in Graham on February 15 and 16, 1877. The Stock Growers' Association of Northwest Texas (forerunner of the Texas and Southwestern Stock Growers' Association) was formed to establish protection against rustlers and a systematic schedule for spring and fall roundups.

Also in 1877 Simpson sold half an interest in all Hash Knife cattle to veteran rancher and banker J.R. Couts, at the same time bringing his brother, Jim Simpson, into the business. The firm of Couts and Simpson hauled lumber in from Weatherford to build a plank ranch house at Hash Knife headquarters. But tracks laid by the Texas and Pacific Railroad in 1880 bisected Hash Knife range. Couts and the Simpson brothers helped organize and name the town of Abilene, beside the railroad about a mile and a half south-west of their headquarters.

These events prompted Couts and Simpson to relocate to uncrowded range in southwest Texas. By 1881, 20,000 Hash Knife cattle were grazing west of the Pecos River below the New Mexico line, with headquarters forty miles north of Toyah. Also in 1881, Dallas banker William E. Hughes paid J.R. Couts $77,500 for his interest in the Hash Knife. Like other Texas ranchers, Simpson began fattening his young steers on northern grasslands, recording the Hash Knife brand in Montana in 1882. This preempted range was in southeastern Montana and across the line in Dakota Territory, and was protected so vigorously by the transplanted Texans that they became known as a "rawhide outfit." Grazing rights were obtained to a large range north of Abilene, and Hughes and Simpson acquired other ranches in North Texas. In 1882 they paid $12 a head for 25,000 cattle (not counting calves) and 1,000 horses from E.B. Millett & Bros. in the largest livestock transaction in the history of northwest Texas.

To facilitate such large-scale expansion, Hughes, Simpson and other investors incorporated the Continental Cattle Company in St. Louis on April 28, 1882. This firm purchased E.B. Millett's big Mill Iron Cattle Company, with eight line camps and a fortified stone warehouse-blockhouse at headquarters near Seymour. (Mill Iron cowboys were a notoriously rough group, readily inclined to use their guns.) The Continental Cattle Company and Mill Iron Cattle Company were reorganized as the Continental Land and Cattle Company at company offices in Dallas on January 21, 1884. William Hughes was named president of the corporation, and John Simpson would continue to serve as manager. The Continental Land and Cattle Company, with over 80,000 head of cattle on three large Texas ranches and a Montana-Dakota range, soon would help establish a still larger property in Arizona.

The Aztec Land and Cattle Company was incorporated in New York City on December 1, 1884. A key figure in founding the Aztec was Henry K. Warren, a native New Yorker who had become a pioneer freighter working out of Weatherford, Texas. His team-sters were quick-triggered men who had to brave Comanche and Kiowa attacks, but Warren prospered, becoming a banker and a director of the Continental Land and Cattle Company — as well as vice-president and general manager of the Aztec at $10,000 per year.

In 1884 Edward Kinsley, a trustee of the Atlantic and Pacific Railroad, inspected the line and was especially captivated by northern Arizona. Aware of the immense profits supposedly available through the Beef Bonanza, Kinsley became excited by the open grasslands that rolled past the train window. From the train Easterner Kinsley could not recognize the arroyos, sink holes, and deep crevices that crisscrossed the countryside, nor could he imagine the lawlessness and violence of Arizona Territory. Back in New York he gathered a group of investors who held controlling interest in the Atlantic and Pacific. The railroad, like other transcontinentals, had been granted millions of acres of government land along the route to help finance construction, and in 1884 the Atlantic and Pacific offered twenty million acres for sale.

The Aztec Land and Cattle Company promptly purchased one million acres in northern Arizona at fifty cents an acre. Railroad land was awarded in alternate sections, checkerboard-style, so that the railroad company would not completely control the right-of-way. But the Aztec was permitted to purchase any sections of its choice, and it chose almost all of the watered sections within the forty-mile limit of the railroad allotment, thereby more than doubling the range under actual control. Aztec range extended from Holbrook on the east to Flagstaff, almost one hundred miles, and about forty miles south of the railroad.

To stock this enormous area, Henry Warren turned to the Hash Knife in Texas. Texas had been scorched by drought, and the Hash Knife agreed to sell "all the cattle and horses and ranch outfit" belonging to its withered trans-Pecos range. Payment would be $50 for each horse and $17.50 per head for all cattle (including calves and yearlings — a high price for the time). There were 2,000 horses and a total of 31,800 cattle, and the Aztec company assumed operation of the Pecos ranch.

The Texas cattle were in poor condition and several trail herds were driven north along the Pecos River to Atlantic and Pacific tracks, then shipped west. About half of the cattle were delivered to Arizona during the summer of 1885. There were heavy losses among the remaining cattle by 1886, but the hardy Texas animals rebounded quickly on the Arizona ranges. Area rancher Will C. Barnes was impressed: "Those Texas cattle could stand more grief, use less food, and bear more calves than any cows that ever wore a brand."

CHAPS, TAPS, AND LATIGO STRAPS

Arizona cowboys were strongly influenced by California *vaqueros* in their attire, tack, and methods of handling cattle. But during the summer of 1885 most of the Texas drovers who delivered 32,000 head of cattle to the Hash Knife stayed on as part of the crew and introduced a new style of cowboying.

"They brought to us the first Rim-fire double-rigged saddles we had ever seen," reminisced rancher Will C. Barnes, a former soldier who, while stationed at Fort Apache, was awarded the Medal of Honor. "We up that way were California center-fire men and used a 70-foot rawhide reata and all that sort of thing. These new boys with their double-cinch saddles, grass ropes, 'tied hard and fast to the nub,' little old potmaker spurs and such new wrinkles surely taught us a whole lot about handling cows. They called us 'Chaps, Taps, and Latigo Straps' in derision of our beloved California outfits. We watched them for a while and then, realizing the unwelcome fact that they knew more cow stuff in a week than we did in a year [,] we quickly put away our cherished high horn single cinch California saddles and 70-foot long reatas, bought a Texas saddle, cached away the long flapping tapaderos we loved so much, bought a 35-foot grass rope and tied it hard and fast, and soon forgot our previous training and accepted the new styles with all good graces."

The original Hash Knife crew in Arizona was made up primarily of the Texas cowboys who accompanied the cattle to Arizona. Many of these Texans were gunmen and fugitives. An unusual percentage gave "Smith" as a last name, and more than a few offered only a first name or a nickname, such as "Poker Bill," "Billy St. Joe," "Ace of Diamonds," and "Loco Tom Lucky." Numerous crew members branded company calves as their own, while many cowboys quit the ranch and regularly stole livestock from the huge herd. Comparing the hard-bitten "Hashknife outfit" with the tough Texas cattle, Will C. Barnes described them as "men of equal meanness, wildness, and ability to survive most anything in the way of hardships and sheriffs."

Henry Warren continued to hire such men, hoping to ward off rustlers and sheepmen. Mormons and other sheepherders constantly tried to move their flocks onto Hash Knife range. Of course, the Hash Knife owned only alternate sections of land, but other sections were open for public grazing. Hash Knife cattle could freely reach public sections, but other stockmen would have to cross Hash Knife land to reach free grazing. The Hash Knife built eight line cabins along the southern edge of company range, from the Mogollon Rim almost to Holbrook.

Hardcases such as Dave Rudabaugh, of Lincoln County War notoriety, hired on as Hash Knife line riders, and it was of little surprise that several Hash Knife men were involved in the vicious Pleasant Valley War, as well as other violent episodes. In December 1886, for instance, a quarrel over a card game in a Hash Knife line cabin erupted in gunplay, and Texan Frank Ward was fatally wounded by Kid Thomas. Another incident late in 1886 brought several Hash Knife riders into Winslow "on a cowboy hilarity," during which they engaged in "promiscuous shooting and yelping on the public streets of the town." Lawmen put the roisterous cowboys on a train to Holbrook, where they all were fired by Henry Warren. On December 8, 1887, Hash Knife cowboy John Taylor, considered a "holy terror," was shot dead outside a Winslow saloon while shouting threats at someone inside.

The Tewksbury Ranch in Pleasant Valley. A drove of half-wild hogs rooted at the bodies of John Tewksbury and Bill Jacobs, gunned down during an attack on September 2, 1887.

THE PLEASANT VALLEY WAR

Arizona's bloodiest range feud bore the ironic name Pleasant Valley War. Pleasant Valley lies south of the Mogollon Rim, but the violence spilled into Hash Knife country and involved Hash Knife riders. The range war pitted sheepherding Tewksburys against the Grahams, a family of small cattle ranchers, along with allies of each clan, and area rustlers. During 1887 and 1888 there were more than twenty casualties, including four members of a rustling family shot by Sheriff Commodore Perry Owens during a spectacular gunfight in Holbrook on September 4, 1887.

Because Hash Knife cattle had been such a consistent target of rustlers, Jim Simpson informed all of his riders that he intended an all-effort against stock thieves. Any man who did not want to participate in the campaign could draw his time, but only five cowboys took their pay. Three of these reluctant warriors — Tom Tucker, Bob Gillespie, and John Paine — gravitated toward trouble and were wounded on August 10, 1887, in one of the early battles of the Pleasant Valley War. And in one of the final incidents of the war, Hash Knife cowboys Jim Scott and Billy Wilson were visiting horse rancher Jim Stott when a lynch posse arrived. Stott, a friend of Hash Knife and Atlantic and Pacific investor F.A. Ames, was suspected of dealing in stolen horses from his ranch on the edge of the Mogollon Rim. When the posse seized Stott on August 12, 1888, Scott and Wilson were with him, and all three men were hanged.

Many Hash Knife riders, however, were men of reputable behavior, and at several line camps married cowboys were stationed with their families. Indeed, these Hash Knife families were envied by area settlers, because they were regularly visited by a ranch supply wagon. Hash Knife headquarters was southeast of St. Joseph and a few miles west of Holbrook. An office house, grain house, and kitchen-mess hall were built just south of town, and a big Hash Knife brand was painted on the office roof. These frame structures cost a total of $850. Dams also were constructed, to create lakes on the range, and artesian wells were developed. When a two-story

Masonic Hall was built in Holbrook in 1887, the Hash Knife rented the lower floor for $150 per year as additional office space.

Hash Knife cattle continued to be trailed from Texas to Montana, eight herds totaling 21,295 in June and July of 1886. But most of these cattle perished during the "Big Die-Up" of the subsequent winter, and the Montana herd remnants were sold off. In Arizona herd increases were consistently disappointing, and calves were counted — against customary practice — to arrive at totals more acceptable to the board of directors. Year after year, however, expenses in Arizona exceeded beef sales, while the original debt incurred in the purchase of land and cattle remained unreduced, requiring continued interest payments.

In 1887 Henry Warren reduced his annual salary to $6,000, then stepped aside the next year as an unsalaried vice-president. Jim Simpson was brought in from Texas as ranch superintendent at $3,000 per year. Simpson won election to the Arizona Territorial Legislature in 1888, but Warren felt that the distractions of political office far outweighed any benefits to the ranch. The Hash Knife continued to operate in the red, prompting Simpson to resign and move back to Texas at the end of 1889. Thirty-three-year-old John T. Jones, a skilled cowboy, was elevated to the vacancy, and he proved to be a highly capable horseback superintendent.

After company president Edward Kinsley resigned in 1890 (he died the next year), vice-president Henry Warren assumed the position. Although Warren stressed the need for economy in operating the Hash Knife, almost immediately he spent $2,600 to build another ranch headquarters, about two miles west of the original frame structures. The two-story adobe ranch house, located south of the railroad, made an impressive sight for Atlantic and Pacific passengers.

The Hash Knife branded 11,296 calves in 1891, a record for the ranch. But there was an extended drought during the 1890s, and rustlers continued to steal thousands of Hash Knife cattle every year. Hash Knife cowboys fired for hair-branding or mavericking usually retreated to the timbered, mountainous country south of the ranch and continued to prey on the big herd. Only 3,333 calves were branded in 1893, a drop of 8,000 in two years.

By this time Hash Knife saddle horses and bulls were old and in need of replacement. Forced to borrow money for operating expenses, Henry Warren resigned the presidency. Subsequent leadership was undistinguished.

THE BABBITT BROTHERS

The Babbitt brothers of northern Arizona built one of the largest cattle empires in the West, along with a miscellany of other commercial enterprises. There were five brothers: Dave (1858-1929), George (1860-1920), Will (1863-1930), C.J. (1865-1956), and Eddie (1868-1943). The Babbitts launched a successful grocery and hardware business in Cincinnati, Ohio, but by the 1880s the brothers had become interested in the western cattle boom.

With the backing of his brothers, Dave spent six months in 1884 on a scouting trip throughout the West. He returned early in 1886 with Will, intending to establish a ranch in New Mexico. Prices seemed high, however, so the brothers headed into Arizona on the Atlantic and Pacific Railroad. Stopping off at Flagstaff, which was rebuilding after a recent fire, they bought a ranch near town for $17,600, almost all of the capital available to the family. There were 864 head of cattle and nineteen saddle horses, and the outfit was named the CO Bar, for Cincinnati, Ohio. C.J. Babbitt joined his brothers within a few weeks; George sold the Cincinnati business and came out early in 1887; and Eddie moved to Arizona in 1890.

Eddie was elected probate judge and won a seat in the Territorial Senate, but he returned to Cincinnati in 1896 to study law and establish a practice. Dave and George founded businesses in Flagstaff, then formed the Babbitt Brothers Trading Company in 1889. Beginning with lumber and hardware, they diversified into ice, reservation trading posts, hotels, banks, a mortuary, movie theaters, and automobile agencies. Small ranchers who fell into debt to them sold or surrendered their outfits.

C.J. and Will built up the cattle operation, working on the range with their CO Bar herd, then investigating other ranches for acquisition. Cattle and sheep were acquired in large and small numbers, along with large and small ranch properties. In 1901 the Babbitts bought remnants of the Hash Knife herd and the famous brand. During the early twentieth century, Babbitt ranches covered more than three million acres, in Arizona and other western states. But the market collapse staggered the Babbitt ranching empire, and a number of the best properties were sold to maintain solvency. Today, however, Babbitt descendants continue to run four cattle ranches and assorted businesses.

A continuing problem that plagued the Hash Knife concerned title to the one million acres that had been purchased. Because the Atlantic and Pacific lands had only been partially surveyed, just over 572,000 acres were deeded to the Hash Knife. The remaining 427,000 acres could not be delivered because they had never been surveyed. In 1895 the railroad agreed to pay $354,451.28 within two years, a sum which included interest and which was secured by 427,000 acres of land. But in 1897 the Atlantic and Pacific Railroad Company, which had never recovered from the Panic of 1893, was sold under foreclosure, and the Hash Knife claim was unprotected by the foreclosure agreement. Having lost 427,000 acres or

Hash Knife superintendent Burt Mossman later served as first captain of the Arizona Rangers, then returned to ranching as manager of the vast Diamond A, with his permanent home in Roswell, New Mexico.
— Courtesy Arizona Historical Society

$354,451.28, and with its herd dwindling because of drought and rustlers, Hash Knife directors decided to liquidate.

John T. Jones, the only effective member of the management team, resigned as ranch superintendent late in 1897. He was replaced by thirty-year-old Burt Mossman, who had managed a ranch north of Phoenix for four years and who was hired for $100 per month. Mossman moved into the southwest corner bedroom on the second floor of the big adobe ranch house. Intending to confront rustlers, he obtained appointment as a deputy sheriff and he purchased three 1895 Winchesters, the first lever-action repeaters to use a box magazine. (Mossman helped organize the Arizona Rangers in 1901, serving as first captain and arming the force with 1895 Winchesters.)

Mossman's principal activity was to sell off the Hash Knife herd. Although there were sales in 1898, much of the year was spent

After leaving Arizona, Burt Mossman soon re-entered the cattle business at Roswell, making his home at this house.

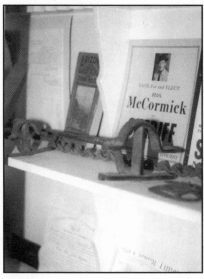

A Hash Knife chuckwagon, displayed by the Navaho County Historical Society at their museum in Holbrook.

A Hash Knife branding iron, displayed by the Navaho County Historical Museum.

in gathering the herd. During 1899, 19,087 cattle were sold at an average of $15 per head (in 1884 the ranch was stocked with Texas cattle at $17.50 per head). In 1900 5,752 cattle were sold at $11.53 per head, and Mossman left at the end of the year.

For a year or so crews rounded up wild cattle for sale, but by 1902 even the horses and saddles were sold, along with the famous brand. Company stockholders still retained the land, however, and tracts began to be leased at 2.8 cents an acre. Large tracts were sold: the sale of 94,855 acres in 1955 brought $4,636,666.96 to be distributed among the stockholders. From 1929 through 1969, the company's land leases and sales were soundly managed by Thomas W. Cabeen of Albuquerque. And today, although little excitement is provided by shootouts or range wars or wholesale rustling, over 200,000 acres of the old Hash Knife range still is leased out for grazing. The Hash Knife brand was purchased in Arizona by the Babbitt brothers of Flagstaff, but was discontinued around 1940; in Montana the brand was reregistered in Carter County and still is used on a ranch owned by descendants of old-time Hash Knife cowboy Frank Castleberry; and in Texas, where it started, the brand is run on part of the old ranch in Baylor and Throckmorton counties.

⇗ SAN BERNARDINO

John Slaughter's famous ranch began as a Mexican land grant issued in 1822 to Lt. Ignacio de Perez. For centuries the San Bernardino Valley was traversed by Indians, Spanish explorers and Jesuit missionaries, and in 1775 the Presidio San Bernardino was built about one mile south of Slaughter's future residence. Perez paid ninety pesos and fees for four *sitios de ganado mayor* (73,240 acres) then erected a large adobe *hacienda* on the site of the old presidio, which had been abandoned in 1780. There was a vast cattle herd, but Apache depredations forced Perez to abandon his *hacienda* and livestock in the mid-1830s. In 1854 the Gadsden Purchase placed the upper third of the San Bernardino Grant above the United States boundary, and U.S. Army detachments often camped at the San Bernardino Springs. But as the decades passed the old *hacienda* buildings fell into total disrepair.

In 1884 John Slaughter, who had been ranching in Arizona for five years, acquired about 65,000 acres of the San Bernardino Grant from a Mexican citizen named G. Andrade, heir of the original owner. The southern two-thirds of Slaughter's acreage stretched into Sonora, Mexico. On the north his range began at the watershed of the San Simon Valley, and the entire ranch was surrounded by mountains. Two streams ran through the San Bernardino, natural springs bubbled in the northern section, and there were numerous cottonwood trees.

"Texas" John Slaughter
— Courtesy Arizona Historical Society

The man who would transform this promising property into a superb cattle ranch was born in Louisiana in 1841. But when John Slaughter was only a few months old his family moved to a land grant in the Republic of Texas. Reared on a cattle ranch near Lockhart, the diminutive John became a fine horseman, a skilled cowboy, and an excellent marksman. John's older brothers, Billy and Charley, also became ranchers, and so did his cousin, Christopher Columbus Slaughter.

John began fighting Indians during the Civil War as a "Minute Man of the Texas Rangers," and he continued to serve in this capacity during Indian alerts through the 1870s. After the Civil War, Slaughter bossed a number of trail drives and established his own ranch in Atascosa and Frio counties. In 1871 he married Eliza Adeline Harris; they had four children, but only Addie and Willie survived infancy.

Charley Slaughter shifted his operation to southeastern New Mexico, where in 1876 John was intercepted on John Chisum's nearby range by rustler Barney Gallagher. Gallagher wanted Slaughter's money belt, but John fatally wounded the badman before he could bring his shotgun into play.

By 1878 John had decided to move his base to Arizona, utilizing Charley's ranch to relay his herds. From Tucson John sent for his wife, little girl and baby boy, but by the time they reached Arizona, they were suffering from smallpox. Although the children survived, Eliza died.

Soon the thirty-seven-year-old widower returned to New Mexico to pick up his second herd from Charley. In New Mexico he became friendly with Amazon Howell and his family. Like Slaughter, Howell was a rancher and Confederate veteran, but John was more interested in his eighteen-year-old daughter, Cora Viola. John and Viola were married in Tularosa on April 17, 1879, then moved to Arizona and established a ranch on the San Pedro River north of Hereford.

By 1883, however, Slaughter was suffering lung difficulties, probably a combination of asthma and a light case of tuberculosis, and he decided to migrate to Oregon. He sold his Arizona herd, but as the stagecoach progressed through Idaho, Slaughter endured a lung hemorrhage. He and his family headed back to Arizona, and at Silver City, New Mexico, Slaughter contracted to supply beef to the Santa Fe Railroad, which made the search for a new ranch site imperative. Soon one of Slaughter's top hands, Tad Roland, described the old San Bernardino grant to John, and the purchase was quickly accomplished.

A compound of adobe buildings went up: at least two dwellings, a stable, smokehouse, milkhouse, and other structures. Viola's parents were installed in one of the houses. Amazon Howell had used a Z brand on the left hip and jaw of his small herds, and Slaughter adopted the same mark, applied to the right shoulder. Howell ran the ranch in the early years, when Slaughter was away acquiring cattle and, after the 1886 election, serving Cochise County as sheriff. Slaughter early had come to a resolute accommodation with the Apaches, and he was known to have dealt violently with rustlers. With lawlessness rampant in Tombstone and Galeyville, Slaughter was pressed into service on the Democratic ticket. He was easily elected in 1886, and submitted to reelection two years later. During his first year in office, Sheriff Slaughter shotgunned murderer Geronimo Baltierrez to death, and in 1888 he killed bandit leader Guadalupe Robles.

Sheriff Slaughter moved his family into Tombstone, while

Viola's parents stayed at the San Bernardino. On March 3, 1887, at 2:13 in the afternoon, an earthquake shook the San Bernardino Valley (tremors were felt as far north as Santa Fe and as far south as the vicinity of Mexico City). The fault line that produced the natural springs in the valley resulted in the earthquake, but after the quake new springs bubbled out of Slaughter's ground. No one was injured at the ranch, but most of the adobe structures were wrecked.

Slaughter refused to run for a third term in 1890, although in 1895 he accepted a permanent appointment as deputy sheriff. After leaving the sheriff's office, Slaughter moved his family to the San Bernardino Ranch, which he would personally supervise for the remaining three decades of his life.

The Slaughters settled into a small house a couple of hundred yards above the international boundary, at the base of the Mesa de la Avanzada. A short distance to the west the Slaughters planned a new headquarters compound, and in 1893 the family moved into a six-room adobe home. There were five bedrooms and a dining room, divided by a hallway that ran east and west, with a porch across the south side. Soon an addition was extended to the west, greatly expanding the dining/living room. The family bathed at a hot spring about three-quarters of a mile from the house (when a bathroom was added, water had to be heated outdoors and brought to the tub, because there was no running water inside). A wash house was erected nearby, along with a commissary, bunkhouses, and a big barn. A two-room adobe school went up east of the house, and when, in 1903, an ice plant opened in Douglas, an icehouse was built just outside the kitchen door. Three hundred pounds of ice at a time were brought to the ranch, where ice cream was served each evening.

Other substantial improvements capitalized upon the natural water supply. Slaughter developed a dozen artesian wells and placed 500 acres of land under irrigation, principally for hay, grain, and vegetables (this land was worked by Mexican sharecroppers, who lived on his Sonora section). Just east of the headquarters compound Slaughter built a rock dam that created a pond which covered more than an acre, and a century later the spring-fed oasis still receives sixty gallons of water per minute.

Slaughter's springs saved him during the severe drought of 1892 and 1893, but during these years he had to mortgage over 8,000 head of cattle and 951 horses in order to raise cash. These loans, totaling

The entrance to Slaughter's San Bernardino Ranch is reached by a gravel road fifteen miles east of Douglas.

The stone building north of the ranch was erected to store grain and feed.

The east wing of Slaughter's home (at right) housed five bedrooms.

West of the Slaughter home was the ice-house, and further west was the water tank. This is the area where Jesse Fisher was killed in 1921.

Slaughter ordered this window from a Sears-Roebuck catalog to enhance his expanded living-dining room.

In 1896 Slaughter acquired the Army Signal Corps telegraph line which had been strung from the ranch to Tombstone. This telephone was mounted in the hallway, next to John and Viola's bedroom.

Viola Slaughter nursed everyone on the ranch, and these shelves in the bathroom held her medicine bottles.

This Sears-Roebuck china cabinet was built into the west wall of the dining room.

SAN BERNARDINO
CREW

John Slaughter ("Don Juan")— Owner and manager, 1884-1922.

Viola Howell Slaughter— The first lady of San Bernardino was nearly two decades younger than her husband, but she reared his son and daughter as well as a host of foster children. Viola was nurse to family and crew, presided over a lively social life at the ranch, and frequently accompanied her husband on cattle-buying trips. Following his death Viola resided in Douglas until 1941, when she died at the age of eighty.

Old Bat (John Battavia Hinnaut)— A former slave from Louisiana, Old Bat became a skilled cowboy while working on Billy Slaughter's Ranch. Old Bat helped trail John's second herd to Arizona, and the two men became mutually devoted. Old Bat often traveled with John on cattle-buying trips as a guard and companion, and he sometimes wore the money belt. When John was absent from the ranch, Mrs. Slaughter usually asked Old Bat to sleep on a cot in the hallway outside her door as a guard. Old Bat was fond of candy, and Slaughter bought him soft drinks by the case. He always carried a toothbrush in his shirt pocket, he played a fife frequently if inexpertly, and he kept a hand-cranked phonograph beside him on the wagon seat. Old Bat died at the ranch in 1921, having served Slaughter for forty-three years.

John Swain ("Sweeney" or "Little John")— Another hand born a slave, "Little John" was about the same age as Slaughter and had been "given" to him when they were youths. By the time Slaughter moved to Arizona, "Sweeney" was a crack shot and a superb tracker who helped his boss out of more than one tight situation with Apaches and bandidos. He later moved to Tombstone to work as a miner, returned to cowboying for a time, served as janitor in the courthouse, and almost reached 100 before dying in 1945.

Tad Roland — A top hand who helped trail Slaughter's first herd into Arizona; crippled by rheumatism in middle age.

Jimmy Howell — Viola's younger brother was bilingual; he kept books in both English and Spanish, and was invaluable in dealing with Mexican cattlemen.

Continued

Continued

Jesse Fisher — A cousin of Viola Slaughter, Jesse served John as a deputy sheriff as well as ranch foreman for many years. When Douglas was founded in 1902, Fisher left the San Bernardino to open a meat market. In 1907 he married and bought a small ranch near Douglas to serve his meat business. But in the spring of 1921 Slaughter called on his old ranch boss to aid with a roundup, and when Fisher returned to the San Bernardino, he was murdered during a tragic robbery at the ranch.

Mae —The most notable of a succession of Chinese cooks.

Lee — A Chinese gardener who worked on the Mexican side of the San Bernardino, Lee delivered fresh vegetables each week to the headquarters compound, but he had to return across the border on the same day in accordance with immigration laws.

$11,500, were repaid in 1894. Slaughter recovered quickly from the drought and acquired nearby properties. For many years the ranges of Arizona remained largely unfenced, and Slaughter's stock grazed as far north as Galeyville, deep into Sonora on the south, to the Huachuca Mountains in the west, and he kept his remuda in New Mexico to the east. For a time he owned approximately 50,000 cattle, and one year he shipped 10,000 steers to market.

Slaughter maintained a close personal control over his operation. He rode so incessantly that he never unbuckled his spurs from his size six boot, and he went armed to the teeth, carrying his hol-

Originally attracted to the San Bernardino by free-flowing springs, Slaughter built a dam and created a 1¼-acre pond east of his home. Clear water flowed into the pond at more than sixty gallons per minute, providing a water supply and irrigation for fields.

stered .44, a shotgun, and a repeating rifle. On the morning of September 19, 1898, Slaughter spotted a man skulking about the ranch, and soon discovered that the suspicious individual was a wanted thief known as Peg-Leg Finney. Slaughter led four of his men in pursuit and located him asleep under a cottonwood about a mile from the ranch house. Slaughter quietly approached the sleeping man and tossed his Winchester away, but Finney suddenly sat up and leveled a cocked pistol. Instantly, Slaughter whipped up his Marlin rifle and triggered a bullet which tore through Finney's right hand and thudded into his chest. Two of Slaughter's men also fired, and Finney died where he lay. After an inquest, he was buried in the little ranch cemetery below the headquarters compound.

A couple of years later, gambler Little Bob Stevens held up a roulette game in Tombstone, stole a horse, and fled in the direction of the San Bernardino Ranch — where Slaughter intercepted and killed him. The next year Deputy Sheriff Slaughter joined a posse which tracked and killed a hired murderer. In May 1894 Slaughter participated in a pursuit of Apache renegades. After a skirmish with the Indians, Slaughter found a one-year-old girl in the abandoned *rancheria*. He took her home and named her Apache May, after the month in which she was discovered.

Apache May ("Patchy") toddled after Slaughter and became the most beloved of the foster children reared by John and Viola. In addition to John's son and daughter, the Slaughters took in a succession of unfortunate children from every ethnic group of the region. These children were made to feel part of the family, and they were the reason Slaughter built a school and provided a teacher. Everyone had assigned chores, but there were frequent picnics, with daily ice cream, horseback riding, and hunting. Slaughter was often hard-edged in dealing with adults, but he thrived on the affection of the children, and brought back gifts from his trips to town. He was devastated in 1900 when Patchy's dress caught fire at a wash pit in the yard, inflicting fatal burns.

Willie Slaughter, as small in stature as his father and equally combative, was married in 1908 to Rosalie Neweham, a teacher at the ranch school. They had a son, John Horton Slaughter II, but Willie contracted tuberculosis and died in 1911 at the age of thirty-three. Addie Slaughter kept the ranch books until she married Dr. William A. Greene in 1903. There were three children — John

Slaughter, William A., Jr., and Addeline Howell — and Addie lived until 1941.

The Slaughters offered warm hospitality, and through the years noted guests at the San Bernardino included Gen. George Crook, Gen. Nelson Miles, and the colorful Emilio Kosterliszky, head of the Mexican Rurales. Burt Mossman, first captain of the Arizona Rangers, gave Slaughter a trained cutting horse, Buck Pius. Guests often played poker around the huge dining table; Slaughter was an inveterate poker player, and Viola became a bluff artist. There was an organ, a piano, and a wind-up Victrola in the living room, where dances often were held. Fond of cigars and liquor, Slaughter took no chances when Prohibition began: he had barrels of whiskey and jugs of mescal buried around the headquarters compound.

The army frequently had used the San Bernardino as an outpost during Apache alerts, and Slaughter had accompanied cavalry units as a guide. In 1896 he acquired the telegraph line strung to his ranch from Tombstone by the U.S. Army Signal Corps, and two years later he converted it to the valley's only telephone service. (When Slaughter left the ranch in 1921 phone service to the valley ended, not to be reestablished until 1983.) In 1910, with the outbreak of the Mexican Revolution, a large military camp was established on the Mesa de la Avanzada above Slaughter's pond. The Slaughter Ranch Outpost was maintained until 1923, and its ruins still can be visited. In 1915 Pancho Villa's men began to slaughter cattle across the border within sight of the ranch house, but Slaughter angrily rode down and returned with twenty-dollar gold pieces in his saddlebags as payment. (When Villa's men threatened Colonel Bill Greene's holdings in Cananea, the promoter mused, "Well, if I had John Slaughter and his shotgun, there wouldn't be many left.")

Slaughter helped found the border mining town of Douglas in 1900, fifteen miles west of his ranch, and he engaged in banking, real estate, and other business activities there. Democrats persuaded him to run for the territorial assembly in 1906, but he served only one term, preferring to devote full-time attention to his ranch and other business interests. Slaughter bought the first of several automobiles in 1912; he never learned to drive, but he enjoyed being chauffeured, and it facilitated the trip between the ranch and Douglas.

In 1921 Slaughter, now in his eightieth year and suffering from swollen feet, persuaded his old ranch boss, big Jesse Fisher, to conduct the spring roundup. Fisher, Viola's cousin, had married and moved into Douglas, where he bought a meat market and built a large house. Loyally, Fisher responded to Slaughter's call, although he telephoned his wife three times a day from the ranch.

In a tragic twist of fate, two ranch employees, Manuel Garcia and Jose Perez, enlisted Arcadio Chavez and Manuel Rubio in a scheme to loot the cash box and murder everyone at headquarters. On the night of May 4 they found $80 in cash, but when Fisher emerged in back of the house with a flashlight, they shot him three times. Slaughter forted up inside the house and Viola phoned Douglas for help, but the conspirators fled.

The culprits soon were apprehended and imprisoned, but a few days after the murder John and Viola moved to an apartment in Douglas near their daughter. Although Slaughter continued to manage the ranch, he died in Douglas on February 16, 1922, after visiting the San Bernardino.

The ranch was leased after Slaughter's death, and the Sonora section was sold to Dr. Manuel Caldreon. The Arizona section was purchased by Marion Williams in 1936. Williams repaired and preserved the deteriorating buildings, which remained in good condition when the Nature Conservancy deeded the headquarters area to the Floyd Johnson Foundation, an historic preservation organization which has opened Slaughter's old ranch to the public. Today it is possible to tour what Viola Slaughter described as the stage of "all the happiness our work, struggles and play gave us at the San Bernardino."

Arizona Ranching in a Corral

Biggest Ranches:
> Hash Knife (100 miles x 40 miles, Holbrook to Flagstaff and south of A & P RR; 2,000,000 acres and 40,000 cattle)
>
> San Bernardino (50,000 cattle at peak, with considerable range in Sonora)
>
> Empire (1,000,000 acres and 30,000 cattle)
>
> Sierra Bonita (30 miles x 27 miles, 20,000 cattle)

Best Ranch Nickname:
> "Cherry Cow," for the Chiricahua Cattle Company (Three Cs brand on nearly 20,000 cattle in Cochise County)

Best Cowtowns:
> Willcox ("Cattle Capital of the Nation") and Holbrook.

Best Saloon:
> Bucket of Blood, Holbrook (renamed after a shootout in the Cottage Saloon produced permanent bloodstains on the floor)

First Rodeos:
> Payson (ca. 1884)
>
> Prescott (July 4, 1888)

Bloodiest Range War:
> Pleasant Valley War (19 killed and six wounded in 1887 and 1888)

Most Famous Fictional Ranch:
> The High Chapparal, owned by Big John Cannon (Leif Ericson) on television (1967-71)

Most Famous Western Star Killed in Arizona:
> Tom Mix (on October 12, 1940, the 60-year-old Mix died of a broken neck when his car went off a road under construction near Florence, west of Phoenix)

10

Dakotas

∿ CHATEAU DE MORES

The open-range cattle boom of the Old West attracted ambitious men with dreams of empire; many were Easterners or Europeans, investing Eastern or European capital to develop their hopes and ambitions. No one dreamed bigger than Antoine Amedee Marie Vincent Amat Manca de Vallombrosa, the Marquis de Mores, a Frenchman who pumped vast sums of Eastern money into a plan of sweeping enormity.

Born in 1858 and trained to be a cavalry officer, the Marquis was highly intelligent and energetic. Bored by army routine in peacetime, he resigned his commission and, on a trip to Paris, met Medora von Hoffman, daughter of a wealthy New York banker. The von Hoffmans were vacationing in France, and the Marquis courted and married Medora in 1882. The family returned to New York, where the Marquis accepted a position in the bank. But a cousin excited the Marquis with tales of a hunting trip to the Dakota Badlands, and the budding frontier visionary interested his father-in-law in an investment centered around the era's fabled Beef Bonanza.

The Marquis discovered an ideal townsite for his plans where the Northern Pacific Railroad crossed the Little Missouri River, and in April 1883 he smashed a bottle of wine over the first tent, christening the headquarters of his dream empire "Medora," in honor of

289

RANCHING IN THE DAKOTAS

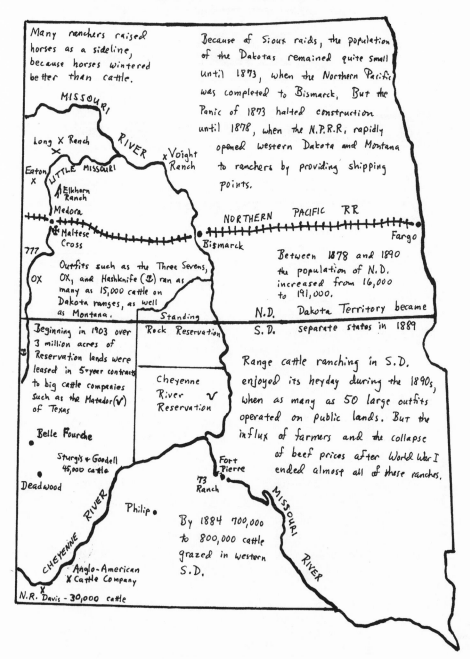

Many ranchers raised horses as a sideline, because horses wintered better than cattle.

MISSOURI RIVER

Long X Ranch X

x Voight Ranch

Eaton X   LITTLE MISSOURI

↑ Elkhorn Ranch

Medora

Maltese Cross

777

OX

Outfits such as the Three Sevens, OX, and Hashknife (⊥) ran as many as 15,000 cattle on Dakota ranges, as well as Montana.

Because of Sioux raids, the population of the Dakotas remained quite small until 1873, when the Northern Pacific was completed to Bismarck. But the Panic of 1873 halted construction until 1878, when the N.P.R.R. rapidly opened western Dakota and Montana to ranchers by providing shipping points.

NORTHERN PACIFIC RR

Bismarck   Fargo

Between 1878 and 1890 the population of N.D. increased from 16,000 to 191,000.

N.D.   Dakota Territory became

Standing

Rock Reservation   S.D.   separate states in 1889

Beginning in 1903 over 3 million acres of Reservation lands were leased in 5-year contracts to big cattle companies such as the Matador (V) of Texas

Cheyenne River ✓ Reservation

Belle Fourche

Sturgis + Goodell 45,000 cattle

Deadwood

Range cattle ranching in S.D. enjoyed its heyday during the 1890s, when as many as 50 large outfits operated on public lands. But the influx of farmers and the collapse of beef prices after World War I ended almost all of these ranches.

Fort Pierre

'73 Ranch

Philip •

CHEYENNE RIVER

By 1884 700,000 to 800,000 cattle grazed in western S.D.

MISSOURI RIVER

Anglo-American X Cattle Company

N.R. Davis - 30,000 cattle

his vivacious and headstrong wife. The Marquis established the Northern Pacific Refrigerator Car Company and built an immense slaughterhouse on the outskirts of Medora, intending to process cattle from area ranches (including his own) and ship dressed meat to the East by refrigerator car. The Marquis also organized a chain of icehouses and cold-storage plants along the Northern Pacific.

The Marquis believed that slaughtering on the range was preferable to shipping cattle alive to distant slaughtering points because shrinkage would be eliminated and the meat would arrive in better condition. This system would permit the western shipper to compete favorably with the eastern shipper of corn-fed beef, while the elimination of middlemen would reduce the price of meat to the consumer. In addition to contracting for thousands of head of beeves from area ranchers, the Marquis immediately began developing his own cattle herd. Within a few months he had 1,500 head, and during his first winter as a rancher (1883-84) he lost just two animals. During the harsh winter of 1886-87 the Marquis again suf-

The Marquis de Mores in his caval-
ry uniform.
— Courtesy State Historical
Society of North Dakota

The Marquise de Mores and her
children, about five years after the
1896 death of her husband.
— Courtesy State Historical
Society of North Dakota

THE NORTH DAKOTA BADLANDS

The plains region west of the Missouri River in North Dakota is an excellent livestock grazing area. Especially favorable ranching conditions prevail in the Badlands of the Little Missouri River, stretching about 190 miles in length and from six to twenty miles in width. Well-watered and grassy, this rough, beautiful land also offers winter shelter to livestock within the multi-colored formation carved by wind and water — buttes, domes, pyramids, and cones.

When A.C. Huidekoper came to the Badlands in 1881, the sole occupants were trappers, buffalo hunters, and wandering Indians. Huidekoper brought the first cattle into the Badlands in 1883, forming the first ranch on the Little Missouri, the Custer Trail Cattle Company, along with the four Eaton brothers. In that same year Theodore Roosevelt and the Marquis de Mores arrived, soon followed by such Texas outfits as the OX, the Three Sevens, and the Hash Knife.

fered minimal losses, primarily because his steers were husky and mature, and were native to Montana and Dakota. The Marquis also raised horses, and his company established a sheep ranch just north of the slaughterhouse.

On a bluff across the river from Medora, the Marquis built "The Chateau," a rambling, two-story ranch house with twenty-six rooms and luxurious appointments. In town the Marquis constructed a two-story brick business "block," a brick church for his wife, and a large brick house to accommodate his in-laws when they visited. The Marquis opened a freight route to Deadwood, a stagecoach line, and a company to ship salmon from the Pacific Northwest to New York City.

Bringing twenty to twenty-five trunks, the Marquis arrived at Medora each spring and stayed until November. The Marquis and Marquise, who was as good a shot as her husband, staged elaborate hunting expeditions. (On one hunting trip into the Big Horn

The Chateau de Mores was completed early in 1884.

Just south of the Chateau were built, left to right, a four-room coachman's house, a stable, and a carriage house.

The meat-packing plant burned in 1907, but the foundations and towering smokestack still stand.

The dinner set of the Marquise comprised 250 pieces of rare china, in addition to silver and exquisite cutlery.

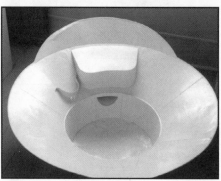

The Chateau boasted the first indoor bathroom in the region; the tub cost $8.75 in St. Paul.

The Marquis de Mores kept a well-stocked wine cellar.

When cowboys helped the Marquise onto one of her sidesaddles, they were required to turn their heads and shut their eyes.

Mountains, she killed three bears, including a grizzly.) Both were excellent riders, but they often used a special hunting coach with their guests. At the Chateau they entertained her parents, rancher Theodore Roosevelt, and other interesting visitors. A French chef prepared ten-course meals which lasted eight hours. But many Westerners were offended by the air of nobility exuded by the Marquis; when challenged by a trio of drunken ruffians, he shot one to death, then won acquittal during a trial which excited enormous local interest.

The Marquis was an active member of the Little Missouri Stockmen's Association, and, since his cattle ranged into eastern Montana, of the Montana Stockgrowers Association. At the 1884 meeting of the latter organization held in Miles City, the Marquis and Roosevelt called for an all-out war on rustlers. The association took no official action, and the Marquis and Roosevelt seem to have taken no part in the subsequent vigilante assaults which resulted in the deaths of at least nineteen rustlers in June and July. But in August the Marquis employed the National Detective Agency of New York to investigate stock theft on his range. Accomplishing several arrests, the detectives learned that he had few real friends. In June 1885, when six of his horses were stolen, the Marquis offered $20 per head for their return, and $500 for the delivery of the rustler or rustlers to Medora.

The infusion of capital in and around Medora stimulated other construction, and by the end of 1884 a population of 250 enjoyed three hotels and a dozen stores. But the youthful Marquis was not an experienced businessman, and his schemes proved overly ambitious and unsound, costing him and his father-in-law well over a million dollars. By 1886 all of his ventures had failed.

The Marquis returned to France the next year, inevitably became embroiled in political controversies, fought several duels, and was slain in North Africa in 1896 at the age of thirty-eight (ambushed by natives, he went down fighting, killing a large number of his attackers). The Marquise raised her three children in France, but returned occasionally to the Chateau. She died in 1921 from the effects of an unhealed leg wound suffered while she served as a nurse during World War I. A caretaker supervised the Chateau for half a century, operating it as a boardinghouse from 1922 to 1936. In 1936 the de Mores family donated the Chateau and other property in the village to the State of North Dakota, and today Medora is a popular tourist site.

✠ MALTESE CROSS AND ELKHORN RANCHES

"The best days of ranching are over," lamented Theodore Roosevelt a year after the disastrous winter of 1886-87. "The great free ranches, with their barbarous, picturesque, and curiously fascinating surroundings, mark a primitive stage of existence . . .; and we who have felt the charm of the life, and have exulted in its bold, restless freedom, will not only regret its passing for our own sakes only, but must also feel real sorrow that those who come after us are not to see, as we have seen, what is perhaps the pleasantest, healthiest, and most exciting phase of American existence."

Among the legions of people who were captivated by the West was the most remarkable American of his age, the dynamic future Rough Rider and U.S. president, Teddy Roosevelt. As an avid outdoorsman and hunter since boyhood, his hopes to shoot a buffalo "while there were still buffalo to shoot" brought him to the Dakota Badlands late in the summer of 1883. He rode the Northern Pacific to the crossing of the Little Missouri River, where a hunting camp for Eastern sportsmen had been established at an abandoned military post. The camp was located on the western side of the Little Missouri, just across the river from Medora, the community recently founded by the Marquis de Mores as the center of his ambitious enterprises. The twenty-four-year-old Roosevelt reveled in the rugged vastness of the terrain, and by the time he finally bagged his buffalo he had decided to join many other upper-class Easterners in becoming a "ranchman."

Roosevelt invested in the ranching enterprise of two Canadians, Sylvane Ferris (his brother, Joe, was Roosevelt's hunting guide) and Bill Merrifield, who had built a log cabin on the Little Missouri about eight miles south of Medora. Roosevelt provided a check for $14,000 to purchase 450 cattle, while Ferris and Merrifield agreed to run the ranch for seven years, receiving half of the herd's increase for their efforts.

The ranch was known as the Chimney Butte, from its location,

Roosevelt's fringed buckskin suit cost $100; his ivory-handled Colt was plated with silver and gold; his silver spurs had his initials; his silver belt buckle had an engraved bear; and his silver-mounted bowie knife was from Tiffany's.
— Courtesy Roosevelt Nature and History Association,
Medora, North Dakota

or the Maltese Cross, from its brand. Cowboys, assuming that "Maltese" was the plural of something, insisted on calling the outfit the "Maltee." Within two weeks Roosevelt, a precocious member of the New York State Legislature, returned to his home.

Tragically, on February 14, 1884, Roosevelt's mother and lovely wife, who had just given birth, died within hours of each other. Hungering for solitude, a grieving Roosevelt returned to the Maltese Cross in June 1884, then decided to establish a second ranch in the isolated Badlands to the north. He brought along Bill Sewall and Sewall's nephew, Will Dow, who formerly had served him as hunting guides in Maine. Roosevelt picked a headquarters site for his new Elkhorn Ranch in a cottonwood grove beside the Little Missouri thirty miles north of Medora and ten or fifteen miles from any neighbor. He sought the privacy to write, although he plunged headlong into the rancher's life, investing another

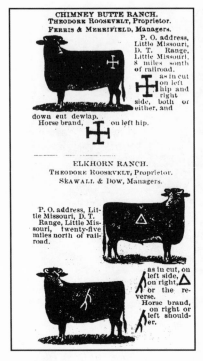

Brand book illustration for Roosevelt's ranches.

$26,000 for cattle and running his herds to 1,600 head. And Roosevelt always would make time for hunting trips, once spending two months on an expedition into Wyoming's Big Horn Mountains. "I work two days out of three at my book or papers," he wrote a sister, "and I hunt, ride or lead the wild, half adventurous life of a ranchman all through it."

Roosevelt worked hard, once riding forty consecutive hours during a roundup and wearing out five horses. Although he was not a good hand with a rope and was only an average horseman, he possessed enormous stamina and courage, willingly riding night herd and performing any other task before him. But Roosevelt, whose maternal relatives were Southerners, regarded the ranchman of the West as akin to a plantation owner of the Old South. He spent considerable time reading and writing and hunting, as befitted a gentleman, and his boots, spurs, guns and clothing were expensive and made to order. He insisted that ranch employees call him Mr. Roosevelt, never Theodore — although out of his hearing he was generally known as "Old Four-Eyes." He good-naturedly shrugged off the frequent remarks about his spectacles, until a drunken ruffian belligerently called him "Four-Eyes" and waved a brace of revolvers for emphasis. Roosevelt, a trained boxer, knocked his tormentor senseless with a right-left-right combination, and thereafter there was little hazing about his eyeglasses.

In October 1884, after Roosevelt had departed to spend the winter in the East, Bill Sewall and Will Dow began constructing the main building of the Elkhorn Ranch, an eight-room log cabin, sixty feet long and thirty feet wide. There was a porch where Roosevelt

"HASTEN FORWARD QUICKLY THERE!"

Roosevelt was regarded by veteran cowboys as a novel specimen because he wore spectacles and gaudy clothing, and because of his speech mannerisms. He seldom used profanity, but his high-pitched voice often piped out head-turning Eastern phrases.

During the spring roundup of 1884, Roosevelt created a Badlands classic when he snapped an order to one of the men: "Hasten forward quickly there!" All cowboys within earshot reeled in their saddles with laughter.

The story was endlessly repeated in bunkhouses and saloons and around chuckwagons. And Medora bartenders patiently indulged thirsty cowboys who asked them to "hasten quickly" with the drinks.

liked to read and write in a rocking chair, and a cellar which he used as a darkroom to process his photographs. A long, low log stable also was erected.

When Roosevelt returned in the spring of 1885 he spent another $12,500 for 1,500 more cattle, placing 1,000 head at the Elkhorn Ranch and 500 on the Maltese Cross. At this point, with 3,350 cattle and 1,100 calves, Roosevelt was the fourth largest cattleman in Billings County. In addition to $52,500 in cattle purchases, Roosevelt expended another $30,000 on wages, equipment, travel, and his cabin. But, like any other Badlands rancher besides the Marquis de Mores, he bought no land, running cattle on "his" preempted range. Roosevelt also owned about eighty horses, including his favorite mount, Manitou. Since his cattle grazed across the line in Montana, Roosevelt joined the Montana Stockgrowers Association, and he founded and served as president of the Little Missouri Stockmen's Association.

One of Roosevelt's employees was "Hell Roaring" Bill Jones, who also served as Billings County sheriff. Sheriff Jones appointed his boss a deputy sheriff late in 1885, and the following March Deputy Roosevelt tracked down and arrested three thieves who stole a

Model of Roosevelt's Elkhorn Ranch cabin, on display at the Visitor Center.

The Elkhorn Ranch cabin and stable, drawn by Frederic Remington.

The Maltese Cross cabin has been moved to the Theodore Roosevelt National Park Visitor Center at Medora.

boat from the Elkhorn Ranch. Although Roosevelt and the Marquis de Mores worked together in stockmen's associations, the two touchy aristocrats clashed on one occasion in September 1885. There was an exchange of letters suggesting a formal duel, but the issue was not pressed by either party.

During the summer of 1885, Will Dow went to Maine to marry his sweetheart, and when he returned he brought his bride, along with Bill Sewall's wife and daughter. "We were all a happy family," wrote Sewall. But a year later, in September 1886, Sewall and Dow decided to take their families back to Maine. The range suffered from drought and overstocking, and cattle prices had plunged drastically. They "squared accounts with Roosevelt," then headed east. Roosevelt already had left (during his years as a rancher he never spent more than a few months without taking the train back to New York). Thus they all missed the devastating winter of 1886-87 on the northern plains.

Roosevelt received the Republican nomination for mayor of New York City. Although he lost the November election, in December 1886 he married a childhood sweetheart, Edith Carow. Following a European honeymoon, he headed west in April 1887 to survey the winter damage, riding for three days without seeing a live steer. "For the first time I have been utterly unable to enjoy a visit to my ranch," wrote Roosevelt, who lost well over half of his cattle.

Although Roosevelt became deeply involved in public life and family affairs, he continued to dabble in ranching, trying for more than a decade to recoup his losses. He spent short periods in the Badlands during the late summers of 1887, 1888, 1890, 1892, 1893 and 1896. By the early 1890s he had abandoned the Elkhorn Ranch, centering his activities on the Maltese Cross; however, in 1892 Roosevelt and three investors organized the Elkhorn Ranch Company, with Sylvane Ferris engaged as manager. Roosevelt's cattle holdings, valued at $16,500, were transferred to this company, and he later invested another $10,200. The purchase of 300 cattle in 1892 increased the Maltese Cross herd to 1,000, and there were other small additions two years later. But his herd slowly dwindled, and in 1898, as he prepared to go to Cuba with the Rough Riders, Roosevelt sold his remaining cattle interests to Sylvane Ferris.

The best estimates are that Roosevelt lost about $50,000 during his ranching ventures, which represented a substantial portion

THE ROUGH RIDERS

As early as 1886 Roosevelt, impressed by the horsemanship, courage and skill with firearms demonstrated by cowboys, began to insist that in case of future warfare a cavalry regiment of such men should be formed. He began to speak of leading "harum scarum roughriders" into combat, and when the Spanish-American War erupted in 1898, Roosevelt resigned as assistant secretary of the Navy to help organize a volunteer cavalry regiment. Offered command of the unit, the militarily-inexperienced Roosevelt wisely accepted a commission as lieutenant colonel, relinquishing the colonelcy (to which he later won promotion anyway) to his friend Leonard Wood, who had been awarded the Medal of Honor for exploits against Apaches.

The First Volunteer Cavalry Regiment, composed of cowboys, Eastern college athletes, and frontier notables such as lawmen Chris Madsen and Bucky O'Neill, trained in San Antonio, where the historic Menger Hotel still sports Roosevelt's name above the barroom. Although other regimental sobriquets were suggested — "Teddy's Terrors," "Teddy's Texas Tarantulas," "Teddy's Cowboy Contingent," "Teddy's Riotous Rounders," "Roosevelt's Rough 'Uns" — the First Volunteer Cavalry soon became known as "Roosevelt's Rough Riders." Once in Cuba, of course, the Rough Riders became the most famous outfit of the war. The acclaim earned by Roosevelt promptly vaulted him to high office (he campaigned accompanied by Rough Riders in uniform), and by 1901 he was president.

of his personal fortune. But the intangible gains far outweighed the financial losses. The long hours in the open air toughened and expanded his once slender physique; the fragile health of his early years was replaced by the boundless energy and robust strength he would enjoy throughout the remainder of his life. Roosevelt's concept of the American character was altered by the proud, vigorous Westerners with whom he lived and worked. He exulted in his participation in this colorful phase of the frontier, compulsively writing articles and books about his ranching experiences. Indeed, Roosevelt was the first gifted and widely read chronicler of this fascinating phase of the last West who was a firsthand participant.

Shortly after Roosevelt became president, the word was spread that "the cowboy bunch can come in whenever they want to," and a number of his western cronies were appointed to government jobs. After Sylvane Ferris was denied entry to the White House, Roosevelt indignantly announced: "The next time they don't let you in, Sylvane, you just shoot through the windows."

Roosevelt remained fully aware of the effects of the West on his physical and spiritual well-being, as well as the sudden prominence accorded him because of his leadership of the Rough Riders. "If it had not been for my years in North Dakota," he declared, "I never would have become President of the United States."

DAKOTA RANCHING IN A CORRAL

Notable Ranchers:

Andrew Voight (1867-1939) — German immigrant who came to northern Dakota as a young man and began ranching on borrowed money. Raised Hereford cattle, Percheron horses, and sheep. Donated beef to nearby Sioux mission, and was named "Andrew, Big Heart White Man Can't Say No."

Scotty Philip (1857-1911) — At 16 emigrated from Scotland; worked as a prospector and army scout; wed a Cheyenne woman and had 10 children. Ranched near Pine Ridge Agency, then in Bad River country where a town was named for him. In 1890 Philip established his noted 73 Ranch near Fort Pierre. He ran from 15,000 to 23,000 head of cattle per year, and he maintained a herd of 900 buffalo.

Biggest Cattle Companies:

Sturgis & Goodelll, on Cheyenne River — 45,000 cattle. Anglo-American Cattle Co., in SW Dakota — 34,000 cattle and 1,000 horses. Harry Oerlichs, president and GM; organized 1882, liquidated 1888.

Best Cattle Towns:

Medora, North Dakota
Belle Fourche, South Dakota (in 1894 shipped 4,700 carloads of cattle)

Most Flamboyant Ranchers: (Tie)

Theodore Roosevelt and Marquis de Mores

Nevada

Ⱶ **SPARKS-HARRELL AND
SPARKS-TINNIN RANCHES**

One of the West's most gigantic ranching empires was begun by Jasper Harrell, who became the embodiment of the rugged, fearless pioneer cattleman. Born on a cotton plantation near Augusta, Georgia, in 1830, Harrell was attracted to the California Gold Rush and sailed for San Francisco in 1850. Mining for several years, he made enough money to acquire agricultural property near Visalia.

Harrell raised cattle, cultivated thousands of acres of cereal grains, and invested in real estate. He purchased his first Nevada ranch in 1870, then bought a herd of Texas longhorns to be driven to the new spread in Thousand Springs Creek Valley. Harrell thus began operations in the Great Basin, a vast sagebrush-grassland frontier bordered by the Rocky, Sierra, and Cascade mountains in northeastern Nevada, lower Idaho, and northwestern Utah. (His nephew from Georgia, Louis Harrell, helped drive the first herd from Texas and became a Nevada cowboy. In 1896, while roping wild horses alone, Louis' rawhide lariat snapped, whipping him across the eyes. He rode two days to the nearest depot, at Montello, Nevada, then took a train to Ogden, Utah, where doctors managed to save the sight in one of his eyes.)

Thousand Springs Creek Valley offered excellent range for fall,

305

RANCHING IN NEVADA

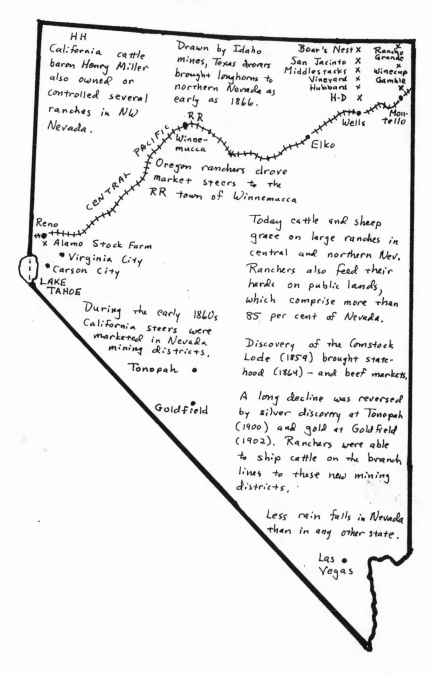

HH
California cattle baron Henry Miller also owned or controlled several ranches in NW Nevada.

Drawn by Idaho mines, Texas drovers brought longhorns to northern Nevada as early as 1866.

Boar's Nest X
San Jacinto X
Middlestacks X
Vineyard X
Hubbard X
H-D X

Rancho Grande X
Winecup X
Gamble X

RR
Winnemucca

Wells
Montello

CENTRAL PACIFIC

Elko

Oregon ranchers drove market steers to the RR town of Winnemucca

Reno
X Alamo Stock Farm
• Virginia City
• Carson City
LAKE TAHOE

During the early 1860's California steers were marketed in Nevada mining districts.

Tonopah •

Goldfield •

Today cattle and sheep graze on large ranches in central and northern Nev. Ranchers also feed their herds on public lands, which comprise more than 85 per cent of Nevada.

Discovery of the Comstock Lode (1859) brought statehood (1864) – and beef markets.

A long decline was reversed by silver discovery at Tonopah (1900) and gold at Goldfield (1902). Ranchers were able to ship cattle on the branch lines to these new mining districts.

Less rain falls in Nevada than in any other state.

Las • Vegas

SPARKS-TINNIN/
SPARKS-HARRELL RANCHES

H-D Ranch — Established by "Old Bill" Downing in the 1860s. Following the emigrant trail to California, Downing stopped off to rest at a spring in the upper region of Thousand Springs Valley. Soon he established a road ranch, selling travelers produce from his large garden and supplies purchased from Wells, to the southwest, and trading for weary and footsore livestock. Downing called his operation the Ox Yoke Ranch but used the H-D brand, which gave the small spread its common name. Located near a major emigrant trail and the Central Pacific Railroad (the closest shipping point was at Wells), the H-D Ranch controlled the head of Thousand Springs Valley and was the first property purchased by Sparks and Tinnin, in 1881. John Sparks and his family made the H-D their home from 1881 through 1885.

Heading north from the *H-D Ranch*, a rider could stop off at the **Hubbard, Vineyard, Middlestacks, San Jacinto,** and **Boar's Nest** ranches, then, across the Idaho line, the **Brown Ranch.** All of these ranches were headquartered west of Salmon Falls Creek. A black foreman, Henry Harris, ran the **Middlestacks** during the 1890s.

Winecup Ranch — Headquartered in eastern Nevada near the Utah line, the Winecup was one of Barley Harrell's favorite ranches. Harrell began irrigating the Winecup in 1875.

Gamble Ranch — South of the Winecup on Thousand Springs Creek and centered near Montello. Today the Winecup-Gamble Ranch runs 12,000 head of cattle on 1,240,000 acres.

Rancho Grande — Located along Goose Creek in the northwest corner of Nevada, the Rancho Grande was a favorite of John Sparks, who began irrigation in 1883.

Shoe Sole Ranch — After James Bower learned of ideal summer grazing in lower Idaho, the Shoe Sole was established by Barley Harrell in three watersheds, the Snake, Humboldt, and Great Salt Lake.

Continued

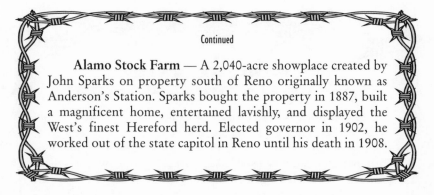

Continued

Alamo Stock Farm — A 2,040-acre showplace created by John Sparks on property south of Reno originally known as Anderson's Station. Sparks bought the property in 1887, built a magnificent home, entertained lavishly, and displayed the West's finest Hereford herd. Elected governor in 1902, he worked out of the state capitol in Reno until his death in 1908.

winter, and spring grazing, but a summer range was necessary at higher elevation. In 1872 Harrell ranch superintendent James Bower was directed to head north with 3,000 head of cattle in search of summer grazing. Two ranchers working a small pioneer spread told Bower about the superb grasslands of the lower Snake River Plains, and soon Harrell bought more Texas cattle and placed them along Goose, Rock and Salmon Falls creeks in northeastern Nevada and south central Idaho. Harrell's cattle wore the shoe sole brand, although his horses and some cattle wore a winecup.

Every three or four months Jasper Harrell personally delivered the payroll to his scattered range crews, fearlessly riding as his own bodyguard. After supper Harrell gambled with his men, often winning back substantial amounts of their wages. Behind his saddle he always carried a grain sack of barley for his horse (his ranches in the San Joaquin Valley produced thousands of bushels of barley), and inevitably he became known as "Barley" Harrell.

In 1872, while Barley Harrell was establishing himself in the Great Basin, traildriver John Sparks delivered a longhorn herd there from the Snyder brothers of Georgetown, Texas, to John Tinnin, a livestock commission agent. The enterprising Sparks soon began to acquire and sell a succession of frontier ranches, often utilizing the services of Tinnin. Sparks and Tinnin moved to Georgetown at about the same time, and in 1881 became partners in the purchase of Bill Downing's H-D Ranch on Thousand Springs Creek in northeastern Nevada. Later in the year Sparks-Tinnin bought a ranch to the east owned by Jasper Harrell and Joseph Armstrong.

In June 1883 Sparks-Tinnin purchased Harrell's other Great Basin ranches, agreeing to a $900,000 price with a $100,000 down-

HONEST JOHN SPARKS

"He was tall, straight as a pine tree, with a clear, flashing eye that seemed to look right through you." Thus did cattleman John Clay describe fellow rancher "Honest John" Sparks. An impressive man who played the role of cattle baron to the hilt and who won back-to-back four-year terms as governor of Nevada, Sparks was the central figure of a ranching empire in Nevada.

The seventh of ten children, John Sparks was born in 1843 in Winston County, Mississippi. The family soon moved to a farm in Arkansas, then to a ranch in Lampasas County, Texas, where fourteen-year-old John quickly learned the cowboy's skills. During the Civil War, John enlisted in a Texas Ranger unit to defend the frontier from Indian depredations. After the war, he joined the great cattle drives out of Texas; and by 1868 he was leading longhorn herds (for "a very small salary," he later grumped) as far as Wyoming, Utah, and, once, Alexandria, Virginia.

In 1872 he married Rachel Knight, the daughter of a physician in Georgetown, Texas. The next year Sparks bought a herd and drove his cattle to Wyoming, establishing a ranch in the Chugwater River Valley and moving his wife to Cheyenne. In 1874 he sold out to A.H. and Thomas Swan for $35,000, then purchased a herd in Colorado and founded a ranch on the North Platte River near Fort Fetterman, Wyoming. Within a year he sold this ranch to a pair of partners from Cheyenne, but promptly founded another North Platte spread, which he soon sold to another Cheyenne rancher. In 1878 he bought the Lodgepole Ranch on the Wyoming-Nebraska border, selling it within a year to a syndicate.

Sparks rapidly pyramided his capital by driving herds to frontier ranges, then selling a series of pioneer ranches to men eager to invest in the western cattle boom. He built a fine home in Georgetown, acquired half-interest in a Georgetown bank, and purchased 10,000 acres in the vicinity. Sparks and his wife had two daughters, but Rachel died in 1879. He married her half-sister, Nancy, the next year, and during the following years they became the parents of four sons. Shortly after his second marriage, he focused his business acumen on the Great Basin, which would lead to wealth and prominence as a cattle baron.

payment and eight annual installments of $100,000, plus four per-
cent interest on the balance. Sparks-Tinnin now owned 80,000-
90,000 head of cattle, and branded 17,000 calves per year. The basic
range ran 150 miles north to south, from the Snake River in southern Idaho to Wells, Nevada, and from Wells on the west about seventy miles east to the Utah line. A great deal of his vast range was
preempted, but at one point during the 1880s it was estimated that
150,000 cattle roamed more than three million acres claimed by
Sparks-Tinnin.

In 1886 Armstrong sold his remaining properties to Sparks-
Tinnin, but that year a long drought set in, severely damaging the
fragile rangelands of the Great Basin. Then the winter of 1889-90
sent thermometers plummeting to -60°, while deep snowdrifts
crusted over, thwarting the desperate efforts of starving cattle and
horses to reach forage on the ground surface. Sparks-Tinnin lost an
estimated 35,000 head of cattle, while the herds of smaller ranchers
were almost totally destroyed. By the spring of 1890 bone pickers
had come to the Great Basin with their mule-drawn wagons, piling
bones along the railroad sidings at Wells, Montello, and Toana.

Tinnin was broke, and after the spring roundup of 1890 he and
Sparks met with Jasper Harrell and his son, A.J., in a log cabin on
Cottonwood Creek south of Twin Falls, Idaho. Sparks had suffi-
cient resources to pay Tinnin $45,000 cash for his interests, then
formed the Sparks-Harrell Company, which incorporated in
California, where the Harrells maintained a home (and a bank) in
Visalia. The capital stock value of $1 million was subscribed equally
by Sparks and the Harrell family.

The disastrous winter of 1889-90 proved the necessity of
growing hay in the Great Basin, but the climate dictated that irrigation be utilized. Jasper Harrell had begun digging irrigation ditches
as early as 1873, soon bringing 3,000 acres under irrigation. When
Harrell rejoined Sparks, the new partnership improved and enlarged
the existing system, irrigating 10,000 acres by 1894. Within a few
years Sparks could boast that his ranches produced 15,000 tons of
hay annually, while his irrigated wildrye meadows (one meadow
alongside Salmon Falls Creek was fourteen miles long) provided the
equivalent of 100,000 tons of standing cured hay. Although cowboys still rode, roped, and branded during the spring and fall, the
tremendous increase in haymaking in the Great Basin focused their

THE ALAMO STOCK FARM

In 1881 John Sparks moved his family from Georgetown, Texas, to the H-D Ranch on Thousand Springs Creek (Sparks continued to maintain and visit his Georgetown home). Although hot springs on the H-D proved beneficial to the health of Mrs. Sparks, the ranch was remote, and in 1885 the family moved west to Reno, primarily to improve educational opportunities for their children. Within two years Sparks purchased 1,640 acres south of Reno, later adding another 400 acres for its irrigated pasture and fertile hay meadow. There was a large grove of cottonwoods, and with his Texas background Sparks knew that the Spanish word for cottonwood is "Alamo." He named his new home the Alamo Stock Farm.

Sparks turned the Alamo Stock Farm into a showplace. He built a handsome plantation-style residence, then funneled the mineral water from two deep artesian wells into a big swimming pool, which allowed Mrs. Sparks to continue to take the waters. Sparks stocked the farm with purebred Hereford breeding stock, and he also brought in buffalo, Persian sheep, elk, and deer. The V&T (Virginia & Truckee) Railroad ran through the Alamo Stock Farm, stopping near his house when needed, and providing passengers with a zoo-like tour as trains rambled across the property.

Sparks proudly displayed his collection of western memorabilia at his home (he paid $100 for a pair of deer antlers locked in mortal combat, but a servant proudly reported to him that after great effort he had finally managed to separate the antlers!). Sparks enjoyed showing off his exotic animals and curios while fortifying guests with his special brew of applejack. Many prominent visitors enjoyed his hospitality, most notably in 1904, when Governor Sparks hosted President Theodore Roosevelt at the Alamo Stock Farm.

summer activities on mucking irrigation ditches and haying and, of course, on feeding hay during the winters.

During the 1890s, Sparks-Harrell invested in alfalfa production, grubbing sagebrush off a designated area, then plowing and discing the new field. It took years to develop productive alfalfa fields, but John Sparks went to the trouble and expense so that registered Hereford bulls could be brought in to upgrade his longhorn herds. The resulting two- to three-year-olds were finished on alfalfa hay, then shipped to California markets.

Sparks relished the role of cattle baron, sporting expensive western attire and carrying a revolver under his coat. A familiar figure at important auctions, Sparks regularly made headlines in livestock journals by purchasing the finest registered breeding stock — most significantly in 1900, when he paid $10,000 for one of the best-known Hereford sires in the national show circuit. Sparks was a founding force in the American Hereford Association, and he became the first rancher to ship purebred Herefords to Honolulu.

While not a technical breeding expert, Sparks employed George Morgan, an Englishman from Herefordshire, to manage his purebred herd at the Alamo Stock Farm, and occasionally to return to Herefordshire to select quality livestock. After showing Theodore Roosevelt around the Alamo Stock Farm in 1904, he introduced the president to George Morgan: "Now that you've met my prize bulls, Mr. President, meet my John Bull."

At the end of each summer Sparks staged a grand deer hunt for important politicians, bankers, and judges from across the West. The guest hunters arrived by train at Wells late in August, when bucks still had antlers in the velvet. The Sparks-Harrell ranges also offered sagegrouse, as well as occasional pronghorns. Sparks provided saddle mounts, a horse wrangler, camp tenders, and a cook, although Sparks himself could prepare a Texas breakfast featuring a baked bull's head, split so that the steaming brains could be ladled onto plates. Sparks hunted with "Alcalde," an old Sharps .50 buffalo rifle that he had notched for bear, buffalo, elk, and deer he had brought down.

During the 1890s John Sparks became increasingly involved in politics, and leadership of the Sparks-Harrell ranches was assumed by Barley's son, A.J. Harrell. Born in 1861, A.J. had attended Heald Business College in San Francisco, then worked with his father in

THE SAGA OF
DIAMONDFIELD JACK

Most of the land grazed by Sparks-Harrell cattle was pre-empted public land, and by the mid-1890s grazing rights were challenged by sheepmen in Cassia County, Idaho. The county's largest ranch was the Sparks-Harrell Shoe Sole. Although Cassia County cattlemen declared a deadline, by 1895 sheepmen had crossed this arbitrary dividing point. John Sparks responded by hiring gunmen, "outside men" who were paid $50 a month, plus bonuses for night riding. They did not perform regular ranch work, but were instructed to keep the sheep back by whatever measures they thought were best. Although admonished to shoot only to wound, they were assured that if killing proved necessary, Sparks-Harrell would provide legal fees and staunch backing.

Of the four "outside men" placed on the Sparks-Harrell payroll, the most dangerous was a moustachioed braggart named Jackson Lee Davis. Because he once prospected in vain for a diamond field, he was known as "Diamondfield Jack." Announcing that he was on "fighting wages," Diamondfield Jack aggressively patrolled the deadline area and threatened sheepherders.

Sheepman Bill Tolman confronted Diamondfield Jack on November 15, 1895, riding up to his Shoe Sole line shack with Winchester at the ready. Diamondfield Jack stepped outside, exchanged a few words with Tolman, then palmed his .45 and drilled the sheepman in the shoulder. To avoid legal repercussions Diamondfield Jack slipped across the border into Nevada, enjoying the cowboy pleasures of Wells, then holing up at the Middlestacks, a Sparks-Harrell ranch.

When Diamondfield Jack left Cassia County, sheepmen soon began to violate the deadline. But Diamondfield Jack moved to the Brown Ranch, a Sparks-Harrell spread a few miles north of the Nevada-Idaho border. On the night of February 1, 1896, he raided a sheep camp, firing a dozen rounds into the camp and killing a horse. The next month two sheepmen, John Wilson and Daniel Cummings, were found shot to death at their camp about fifteen

Continued

Continued

miles north of Brown Ranch. Diamondfield Jack and another Sparks-Harrell gunman, Fred Gleason, promptly vanished and were widely blamed for the killings. Early in 1897 it was learned that Diamondfield Jack was incarcerated in Yuma Territorial Prison for a recent Arizona offense. He was extradited to Idaho, and at about the same time Gleason was arrested in Montana.

Sheepmen raised a large war chest and employed talented special prosecutors. But Sparks-Harrell lived up to the promise to provide every support for their "outside men," spending from $30,000 to $100,000 during the next several years. The defense team was led by James Hawley of Boise, a legendary trial lawyer who would be elected governor of Idaho in 1910.

The trial was held in April 1897 in Albion, the seat of Cassia County. Despite a strong defense case, Diamondfield Jack was pronounced guilty and sentenced to hang. Hawley appealed the conviction, then won acquittal for Gleason. The transcript of Diamondfield Jack's trial was printed for use by the Idaho Supreme Court, but the printing job was awarded to a sheep owner who altered the transcript to favor the prosecution.

Not surprisingly the verdict was upheld, but Hawley was convinced that information was being hidden, and he wrote to John Sparks asking that the real killers be revealed. At last, convinced that an innocent man was about to be executed, James Bower confessed to Hawley. Bower, a longtime Sparks-Harrell foreman, had ridden with cowboy Jeff Gray into the sheep camp. When Bower began scuffling with Wilson, Gray shot both sheepherders. Loyal to their friend and employee, Sparks and Jasper Harrell kept Bower's story a secret, convinced that Diamondfield Jack could not be hanged — since he was innocent.

Bower and Gray made sworn statements, but a stay of execution could not be obtained until the day scheduled for Diamondfield Jack's hanging. Incredibly, his execution was repeatedly rescheduled, and his eighth reprieve arrived only hours before he was to die on July 17, 1901. Despite the confession of Bower and Gray, who claimed self-defense and were not prosecuted, Diamondfield Jack was resentenced to life in prison, and it took Hawley another year and a half to secure a pardon. Although Diamondfield Jack spent six years behind bars, Idaho newspapers railed that freeing him was a miscarriage of justice.

banking and real estate, before immersing himself in the ranching empire. Early in 1901 he took over total control, when Sparks sold out to the Harrells.

Sparks had begun to invest in mining ventures, and the emphasis upon purebred Herefords and irrigated hay fields rendered the ranching enterprise increasingly expensive to conduct. Financial pressures may have forced Sparks to sell his ranching interests, as well as his focus on politics — the following year he made a successful race for Nevada's governorship. Sparks sold 20,000 head of cattle along with a one-half interest in 700,000 acres of rangeland, as well as an additional 700,000 acres of leased land. In return for his half share of Sparks-Harrell, it was thought that Sparks received $500,000 and 12,000 acres of cotton land in Texas. "Cotton is king, you know," remarked Sparks, "and if I raise enough of it, I may make some money — there is no telling."

On May 13, 1901, less than two months after the sale, Barley Harrell died at the age of seventy-one. By this time decades of overgrazing had severely damaged Sparks-Harrell ranges (the company name continued to be used until 1908), and A.J. Harrell began experimenting with grasses from the Russian steppe. But Harrell died at forty-seven in 1908, the same year that Governor Sparks, ill and nearly bankrupt from unfortunate mining investments, passed away at his beloved Alamo Stock Farm.

The Sparks-Harrell ranches soon were purchased by the Utah Construction Company, which maintained ownership through World War II. Today the old ranches are controlled by diverse owners, and much of the range has recovered from the overgrazing of the late nineteenth century. The Middlestacks headquarters today is an abandoned ruin, while the Brown Ranch is deeply submerged beneath what now is the Salmon Creek Reservoir. But several of the old ranches are still used, most notably the Winecup, purchased by Sparks-Tinnin from Joseph Armstrong in 1886, and the Gamble Ranch near Montello.

The Winecup-Gamble maintains two headquarters, one at each ranch, and in most years runs 12,000 head of cattle on 1,240,000 acres of desert rangeland — 290,000 acres of deeded land and 850,000 acres leased from the Bureau of Land Management. Rainfall on the Winecup-Gamble averages three to eight inches per year, forcing reliance upon windmills. Elevations range from 4,800 feet to

10,500-foot mountain peaks. Alfalfa is grown on 1,500 acres of irrigated land. Forty-five ranch hands work through the summer months, and twenty-five are employed year-round. All horses are bred on the ranch. There are six stallions and about 100 brood mares, producing perhaps eighty foals each year. Cattle are primarily Hereford cows bred to Angus and Brangus bulls. Wildlife still abounds on the ranch. The Winecup-Gamble worthily carries on the traditions of a ranching empire established more than a century ago.

NEVADA RANCHING IN A CORRAL

Biggest Ranching Empire:
Sparks-Tinnin/Sparks-Harrell Ranches (From Wells, 150 miles north into Idaho, and 70 miles east into Utah; 150,000 cattle)

Notable Nevada Cattlemen:
John Sparks, who came to Texas to build an enormous ranching empire, and who twice won election as governor of Nevada.

Jasper Harrell, progressive cattleman who bought the first of several Nevada ranches in 1870.

Lewis R. Bradley, who came over the Sierra Nevada with cattle in 1862, then served as second governor, 1871-79.

James Hardin, "Cattle King of the Humboldt Valley."

Henry Miller, California baron who also operated several ranches in NW Nevada and SE Oregon.

W.B. Todhunter, California rancher who expanded into Nevada's Paiute Valley.

Hugh Glenn, Oregon rancher who also expanded into the Paiute Valley.

Most Famous Fictional Ranch:
The Ponderosa, owned by Ben Cartwright (Lorne Green), on the television series *Bonanza* (1959-73).

RANCHING IN CALIFORNIA

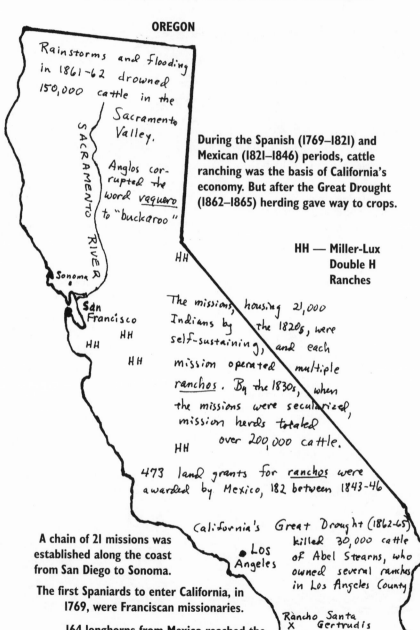

OREGON

Rainstorms and flooding in 1861-62 drowned 150,000 cattle in the Sacramento Valley.

Anglos corrupted the word *vaquero* to "buckaroo"

SACRAMENTO RIVER

Sonoma

San Francisco

HH

HH

HH

HH

HH

During the Spanish (1769–1821) and Mexican (1821–1846) periods, cattle ranching was the basis of California's economy. But after the Great Drought (1862–1865) herding gave way to crops.

HH — Miller-Lux Double H Ranches

The missions, housing 21,000 Indians by the 1820s, were self-sustaining, and each mission operated multiple ranchos. By the 1830s, when the missions were secularized, mission herds totaled over 200,000 cattle.

473 land grants for ranchos were awarded by Mexico, 182 between 1843–46

California's Great Drought (1862-65) killed 30,000 cattle of Abel Stearns, who owned several ranchos in Los Angeles County

● Los Angeles

A chain of 21 missions was established along the coast from San Diego to Sonoma.

The first Spaniards to enter California, in 1769, were Franciscan missionaries.

164 longhorns from Mexico reached the missionaries at San Diego in 1769.

Rancho Santa Gertrudis X
● San Diego

California

RANCHING IN CALIFORNIA

After establishing missions and settlements in Baja California throughout the first half of the eighteenth century, the Spanish moved north into present-day California, founding a mission and presidio at San Diego in 1769. Eventually there would be a chain of twenty-one missions, each about a day's walk from the next mission. Among other agricultural activities, cattle were introduced and raised at the missions. Although there was little market for beef, hides and tallow could be sold.

A *rancho* system was initiated late in the 1700s. A sizable *rancho* would be granted to a *ranchero* family, which would exercise almost complete control over the land and the mission Indians assigned to their land. (Even though some of these grants were quite large, the term *rancho* was used in California to the exclusion of *hacienda*.) When Mexico became independent of Spain in 1821, only a little more than twenty *ranchos* were in existence. But in 1834 the California mission lands were secularized, then granted or sold as *ranchos*. Some of these *ranchos* were stocked with sheep, but most were cattle operations.

Late in the 1700s, retired soldier Manuel Nieto secured a grant of more than 150,000 acres above San Diego, and after his death in

319

1804 his heirs kept the Rancho Santa Gertrudes intact for three more decades before partitioning the property into six ranchos in 1833. Five years later the choicest Nieto division, Rancho Los Alamitos, was purchased by Abel Stearns, an adventurous native of Massachusetts who had become a wealthy Los Angeles merchant. Naturalized as a Mexican citizen, when he was forty "Don Abel" married the four-teen-year-old daughter of an aristocratic family of old Californios. A shrewd businessman with excellent connections, Don Abel pur-chased half a dozen more ranchos in Los Angeles County, accumulat-ing more than 200,000 acres by 1862 and achieving recognition as the most important ranchero in southern California. But California's Great Drought began in 1862, killing 30,000 Stearns cattle, and Don Abel's financial condition became precarious. Unable even to pay his taxes, Don Abel had lost control of his ranchos by the time he died at the age of seventy-three in 1871.

Such ranching empires, large by California standards, were only a fraction of the size of the vast spreads that developed in the Great Plains during the two decades following the Civil War. The Forty-Niner Gold Rush, occurring shortly after the United States wrested California from Mexico, provided mining camp markets for new ranchers from the United States. The most successful of the Gold Rush-era ranchers was Henry Miller, who by the 1880s ran 100,000 cattle on more than a million acres of land in California, Oregon, and Nevada.

While Miller developed his empire, during the 1860s and 1870s, cattle ranching declined overall in California. After the first few years of the Gold Rush, agriculture began a shift from stock raising to intensified farming, a trend accelerated by flooding and drought. Terrible rainstorms in 1861 and 1862 triggered floods which drowned 150,000 cattle in the Sacramento Valley, then the Great Drought of 1862-65 further devastated ranchers. Ranchers irrigated their land and turned to farming or sheep-raising. In 1860 there were about one million sheep in California, but within less than two decades the sheep population had soared to nearly eight million. Wheat production also exploded, and farmers cultivated grapes and oranges as well.

Extolling California's splendid climate, real estate agents stim-ulated a land rush in the 1870s which rivaled the Gold Rush. The state's population doubled to more than 1.2 million by 1890, but by

1900 the average California "ranch" averaged less than 400 acres. In California it became customary to label as a "ranch" twenty-acre property with, say, a vineyard or orchard.

California has a ranching tradition dating back to the Spanish origins of *ranchos* and *vaqueros*. But shortly after Anglo-Americans began to arrive in significant numbers, California cattle ranchers became plagued by disastrous weather conditions and by agricultural opportunities in farming and sheep-raising. At the very time when enormous ranches were taking shape across the heartland of the open ranges of the West, Californians turned toward more promising economic pursuits.

HH   HENRY MILLER AND THE DOUBLE H

"I have made three fortunes," remarked Henry Miller, "one for myself, one for my partner, and one for my lawyers."

The man destined to become California's greatest cattle baron was born Heinrich Alfred Kreiser in Germany in 1827. The son of a butcher, he bought livestock for his father's shop and tended the family herds. He left in 1844, turning up three years later as a butcher in New York City. When a young acquaintance named Henry Miller decided to back out of a planned trip to the California gold fields, Heinrich bought his nontransferrable boat ticket, assuming Miller's name to cash in on a bargain price.

Arriving in San Francisco in 1850, "Henry Miller" found work as a butcher, and within a year opened his own meat market. His chief competitor in San Francisco was an Alsatian-born butcher named Charles Lux. In 1857 Miller formed a partnership with Lux, eliminating a key rival while acquiring access to his excellent financial connections. Needing land to graze the firm's livestock, Miller bought an 8,800-acre ranch in the San Joaquin Valley from two brothers named Hildreth. Their HH brand was included in the transaction, and the Double H would become the primary brand for Miller's ranching empire.

Aggressively and unscrupulously Miller acquired land for thirty years. Some of the lands were irrigated and farmed, and during the 1870s Miller and Lux expanded into the sheep business. By the time Lux died in 1887, there were 100,000 cattle and 80,000 sheep on more than fifteen Miller & Lux ranches, and eventually Miller gained sole ownership of the firm's property from Lux's heirs. Miller owned or controlled more than 800,000 acres in nineteen California counties, as well as another 200,000 acres in southeastern Oregon and northwestern Nevada. Miller also owned banks, hotels, slaughterhouses, stores, and lumberyards, accumulating considerable political clout as his wealth mounted.

THE DIRTY PLATE ROUTE

During the open-range period, a majority of cowboys were unemployed through the winter months. Many out-of-work cowboys rode the grub-line, spending the winter traveling from one bunkhouse to another. The occasional "grub-line rider" (sometimes called a "grub-liner") brought news from the outside and was welcome company at isolated ranches.

But the hospitality extended to grub-liners inevitably attracted tramps. These unwelcome parasites scrounged as many meals as possible, balked at doing chores, and camped at faraway corners of the range, sometimes carelessly allowing their campfires to set the grasslands ablaze. Cattle baron Henry Miller, who owned more than fifteen ranches in California, Nevada, and Oregon, pragmatically issued a set of rules regarding tramps:

1. Never refuse a tramp a meal, but never give him more than one.
2. Never refuse a tramp a night's lodging. Warn him not to use any matches and let him sleep in the barn, but only for one night.
3. Never make a tramp work for his meal. He is too weak before and too lazy afterwards.
4. Never let tramps eat with the men. Make them wait and eat off dirty plates.

Tramps continued to haunt Miller's ranches, but began to term them the "dirty plate route."

"I consider twenty-four hours a working day for me," declared the humorless Miller. "If I get my work done before the twenty-four hours is over, I sleep. But the work must be done, whether I get my rest or not."

Tirelessly and with relentless attention to detail, Miller patrolled his empire in a buggy. One of his finest ranch managers was Ace Mitchell, who superintended his Oregon operations. But Mitchell forgot to kill and butcher a lump-jawed steer as ordered by the boss, and when Miller spotted the steer on a visit two years later, he fired Ace on the spot.

Miller's industry and foresight allowed him to prosper when others suffered disaster. The severe drought of 1888, for example, wiped out many California stockmen. But there was water for his meadows. Cattle that did die were immediately skinned for their hides; their fat was extracted in the rendering vats on each of his ranches, and the remains were made into chicken-feed. Miller then bought out the smaller operators who had gone under, and during the five years after the drought his profits totaled $8 million. Small wonder that when Henry Miller died in 1916, at the age of eighty-nine, accountants tallied the worth of his estate at $40 million.

Ranches to Visit

Western travelers interested in historic ranches may find many tangible remains that are easily accessible to the public, three dozen of which are listed here. Not listed are many working ranches with historic pasts; a polite inquiry and a willingness to traverse back roads might result in a uniquely rewarding travel experience.

TEXAS

King Ranch. The world's most famous ranch provides bus tours seven days a week. There is an admission fee, and tours leave every hour, but there sometimes is a considerable wait. There is a museum at the waiting point, which is on the western outskirts of Kingsville on Santa Gertrudis Avenue. Another King Ranch Museum is in downtown Kingsville, along with the King Ranch Saddle Shop (Richard King began operating his own saddle shop nearly 130 years ago).

XIT. In Channing the brick general office, built in 1890, has been handsomely refurnished and may be viewed at the north end of the little town's main street. In Dalhart an excellent museum focusing on the XIT is across the street from the courthouse, and for decades an annual XIT reunion and rodeo has been staged the first weekend in August. The event is extremely popular, even though not as much a reunion as before. The 97-year-old Ira Taylor (as of this writing) is the only surviving XIT cowboy.

Ranch Heritage Center. Internationally famous museum featuring more than thirty structures — ranch house, bunkhouses, stables, barns, and miscellaneous outbuildings — from historic West Texas ranches. Located in Lubbock adjacent to the campus of Texas Tech University.

Pitchfork. Headquarters is west of Guthrie on the south side of U. S. Highway 82. The ranch maintains a museum in the old two-room office building, and with typical western hospitality guests will be escorted through the headquarters complex.

6666. On the outskirts of Guthrie at the intersection of U.S. Highways 82 and 83 is the stone commissary of the Four Sixes, the oldest building still standing that was erected by Burke Burnett. It is used as a store today, and a traveler may make purchases at the historic old storehouse. Across Highway 82 to the north is the magnificent Four Sixes ranch house and a picturesque barn used in Marlboro Man ads.

Lazy S. A mile south of Morton on the west side of Highway 214 stands the splendid *hacienda*-style headquarters complex of the Lazy S. Built in 1915 by craftsmen brought in from Mexico, the old complex is not open to the public, but inquiry at the Slaughter Ranches office in Morton should result in permission to visit.

Hill Country State Park. Several years ago the State of Texas acquired a 5,400-acre cattle ranch that dates back to the nineteenth century. Located ten miles west of historic Bandera, the old spread now is an equestrian park. Primitive campsites are available, along with trail maps. Visitors bring their horses in trailers and spend days riding through the scenic Hill Country among old ranch buildings and corrals. The park superintendent resides in the 1916 ranch house, but he can furnish just six of the twenty rooms. For those who cannot bring horses, the adjacent Running R guest ranch provides long trail rides through the park.

Cattlemen's Museum and Stockyards Historic Area. In Fort Worth the Southwestern Cattle Raisers' Association provides an excellent museum at 1301 West Seventh. Along Exchange Avenue in Fort Worth's North Side is the atmospheric Stockyards area, featuring the magnificent Stock Exchange building and, across the street, the world's largest mule barn, along with numerous other fine structures from the turn of the century. A few blocks away Burke Burnett, W.D. Waggoner, and other cattle barons rest beneath ostentatious monuments at Oakwood Cemetery.

Dan Waggoner Home. At 1003 E. Main in Decatur stands the two-story limestone mansion built in 1883 by Dan Waggoner for his growing family. Dominating a hill at the end of Main Street, "El Castile" is not open to the public, but the one-time headquarters of Waggoner's ranching empire amply rewards a look from the outside.

Charles Goodnight Home. On his last ranch the legendary cattleman built a handsome frame Victorian house in 1892 which served as his headquarters and home until he died in 1927. Although still occupied today, the Goodnight home may be viewed and photographed just south of U.S. Highway 287 at the hamlet of Goodnight, thirty-three miles east of Amarillo.

OKLAHOMA

Dog Iron Ranch. The two-story log and plank home of Clem Rogers has been moved a short distance from its original site, which now is a lake bottom. Will Rogers was born in the home in 1879. The ranch house, an impressive barn, and other outbuildings are maintained as a state park, two miles east of Oologah.

Pawnee Bill's Ranch. On a hill on the south side of U.S. Highway 64 just west of Pawnee is the headquarters complex of the famous showman. The fourteen-room ranch house was built in

1910, and there are handsome stone buildings, as well as a museum and grazing longhorns and buffalos. Pawnee Bill's Ranch is open to the public.

National Cowboy Hall of Fame and Western Heritage Center. Superb collections on cowboys, ranching, and rodeos await visitors to this internationally famous museum at 1700 Northeast 63rd Street in Oklahoma City.

KANSAS

J. Fulton Pratt Ranch. On the north side of U.S. Highway 24 just west of Studley, the State of Kansas maintains a superb stone ranch complex built by Cornish masons in the style of an English estate.

Z Bar Ranch. On a hill on the west side of Highway 177, just north of the intersection with U.S. Highway 50, stands an imposing three-story stone ranch house built in 1881. South of the house is the equally impressive stone barn of the Z-Bar, and the 11,000-acre ranch has been purchased by the National Park Trust for eventual utilization as a prairie preserve.

NEBRASKA

Scout's Rest Ranch. At the Buffalo Bill Ranch State Historical Park in North Platte, the beautiful ranch house, enormous barn, and numerous outbuildings of Cody's Scout's Rest are open to the public. One of the best ranch tours in the West.

Bowring Sandhills Ranch. Just north of Merriman on Highway 61, take a right to the property of Nebraska rancher-politician Arthur Bowring. Now a state park, the ranch house and outbuildings are open to the public, along with a modern museum and a reconstruction of the original adobe ranch house, currently operated as a living museum to the pioneer ranchers of the vast sandhills country.

COLORADO

White House Ranch Historic Site. Located at 3202 Chambers Way near the entrance to the Garden of the Gods at Colorado Springs, the White House Ranch features an 1895 working ranch, an 1868 homestead, and a 1907 mansion as part of a living history program. Also of interest at 101 Pro Rodeo Drive (I-25 Exit 147) is the Museum of the American Cowboy and Pro Rodeo Hall of Fame.

Two Bar Ranch. The abandoned log buildings and corrals of Ora Haley's Two Bar are in Brown's Park, just south of Highway 318 where the road crosses the Little Snake River. Drive south off 318 to the old schoolhouse on a hill, then hike downhill northwest to the splendid ruin.

WYOMING

Mason-Lovell Ranch. Another lonely log ruin is maintained by the state thirteen miles east of Lovell on the south side of U.S. Highway Alternate 14. Continuing to the east, the traveler will be rewarded with one of the West's most spectacular drives.

Trail End Historic Center. At 400 Clarendon Avenue in Sheridan is the three-story home of John B. Kendrick, who arrived in Wyoming as a Texas cowboy on a trail drive. Intelligent and thrifty, Kendrick built a ranching empire, then was elected governor and U.S. senator. From 1908 through 1913, he erected Trail End as the family home.

Bradford Brinton Memorial Ranch Museum. The ranch was purchased in 1923 by Brinton, who enlarged the 1892 ranch house to twenty rooms. A tour includes outbuildings and an art museum. Drive south of Sheridan seven miles on U.S. 87, then five miles southwest on Highway 335.

TA Ranch. Thirteen miles south of Buffalo to the east of

Highway 196 is the TA Ranch, site of the climactic battle of the Johnson County War. Now operated as a guest ranch, the historic TA features the log barn with numerous bullet holes visible from the loft.

Swan Ranch. The impressive buildings of the vast Swan Land and Cattle Company include the headquarters office and commissary. Exit I-25 at Chugwater, drive west through town, cross the railroad tracks, and find most structures on the right. Buildings are closed, but there are rich photo possibilities, and it is fun to walk onto the porches and peer in the windows.

Wyoming Hereford Ranch. Historic nineteenth century ranch today is the home of a magnificent Hereford herd and a fascinating complex of buildings and corrals. Visitors are welcome; drive three miles southeast of Cheyenne on I-80 and take Wyoming Hereford Exit south.

Montana

Grant-Kohrs Ranch. The National Park Service maintains the headquarters of the million-acre Grant-Kohrs cattle operation as a living history museum of western history. Located at the north edge of Deer Lodge in a beautiful mountain setting. One of the West's most rewarding ranch tours.

Pierre Wibeaux House. At Wibeaux in eastern Montana is the furnished town house and office of the founder of the W-Bar Ranch. Across the street a railroad car serves as a Montana Information Center and a museum where a tour of the Wibeaux house may be obtained. On a hill at the western edge of town stands a statue erected by the cattle baron of himself.

Ranch Riders Museum. On the north side of U.S. Highway 10 at the west end of Miles City is a museum with extensive collections relating the ranching history of Montana.

ARIZONA

San Bernardino Ranch. The headquarters complex of John Slaughter's famous San Bernardino Ranch is preserved almost intact. The adobe ranch house and several other buildings seem ready for Slaughter and his family to return from a trip and resume their busy ranch activities. Drive east from Douglas fifteen miles over a good gravel road.

Empire Ranch. The Bureau of Land Management has preserved the headquarters buildings of Walter Vail's Empire, and currently leases 45,000 acres of the historic ranch. From Sonoita drive north on Highway 83; between mileposts 39 and 40 turn east (gate sign reads "Empire Cienegos Resource Conservation Area"); drive four miles to ranch buildings on the left; tour usually is self-guided.

Pete Kitchen Ranch House. Considerably altered in appearance, Pete Kitchen's fortified adobe, which withstood numerous Apache attacks, stands near U.S. Highway 89 on the northern outskirts of Nogales.

Tanque Verde Ranch. The historic old spread on the western edge of the Rincon Mountains today is a luxurious guest ranch, using several of the most venerable Tanque Verde buildings. Drive east out of Tucson to the end of Tanque Verde Road.

NORTH DAKOTA

Maltese Cross and Elkhorn Ranches. Theodore Roosevelt's first ranch in Dakota Territory was the Maltese Cross, six miles south of present-day Medora. His log cabin has been moved to the western edge of Medora, at the visitor center of the Theodore Roosevelt National Memorial Park. The visitor center museum is devoted to TR's ranching experience, and includes a large model of his Elkhorn Ranch cabin, which no longer exists. The 70,416-acre park stretches north of Medora, and by rough roads the Elkhorn Ranch site may be reached.

Chateau de Mores. On a high plateau southwest of Medora, the twenty-six-room Chateau de Mores overlooks the town built by the Marquis in the 1880s. The house is open daily, along with the stable and other buildings.

Ranching and
the Movies

Cowboys decorate most western films, and are central characters in a majority of oaters. But surprisingly few westerns have
much to do with cowboying and ranches — the predominant sub-
jects are gunfighters and outlaws, or cavalry vs. Indians.

Several western films ostensibly about ranchers or cowboys do
not really focus on the range. John Wayne played a cattle baron in
Chisum, Big Jake, and *McClintock!,* but the Duke spent most of his
time brawling and gunslinging. Although James Cagney, Lionel
Barrymore, and Robert Taylor also played powerful cattlemen in,
respectively, *Tribute to a Bad Man, Duel in the Sun,* and *Cattle King,*
it is difficult to understand how their characters achieved success,
since they spent so little time in ranching activities. Despite on-
screen assistance from co-star Ronald Reagan, Barbara Stanwyck
was only a nominal *Cattle Queen of Montana.* Supposedly a trail-
driving outfit, *The Culpepper Cattle Company* proved to be primarily a gang of outlaws and gunfighters. And the recent hit, *Legends of
the Fall,* had a picturesque ranch in the background, while sex symbol Brad Pitt dominated the foreground.

But Charlton Heston as *Will Penny* and Lee Marvin as *Monte
Walsh* starred in Hollywood's finest films about cowboys and the

332

Kirk Douglas shows fellow cowboy William Campbell the scars from an encounter with barbed wire in Man Without a Star.

ranching frontier. John Wayne was superb in two trail-driving epics, *Red River* and *The Cowboys*. Young, lanky Gary Cooper was believable and appealing in the 1929 version of *The Virginian*, while Kirk Douglas captured the essence of a footloose cowboy battling barbed wire in *Man Without a Star*. Cattle barons and their empires were vividly portrayed by Spencer Tracy in *Broken Lance*, Edward G. Robinson in *The Violent Men*, Charles Bickford in *The Big Country*, Walter Huston in *The Furies*, and Rock Hudson in *Giant*.

Burt Lancaster was dynamic and rugged as a ranch foreman in *Vengeance Valley* and as the head of a beleaguered ranch family in *The Unforgiven*. Maureen O'Hara and James Stewart introduced the first Hereford bull to the West in *The Rare Breed*. There is an interesting look at big Wyoming ranchers in Steve McQueen's *Tom Horn*, and *Ramrod, Cowboy*, and *Ride the Man Down* are tough, gritty cowboy movies. Captivating portraits of post-World II ranch life may be seen in *Comes a Horseman*, with Jane Fonda and James Caan; *Hud*, starring Paul Newman; and *The Rounders*, with Glenn Ford and Henry Fonda sharing the screen with a diabolically rambunctious bronc.

There have been hundreds of western movies and cinematic cowboys. But only rarely has the gunsmoke cleared away long enough to give viewers a true look at ranching and cowboy life.

Rest in Peace

1867 — Oliver Loving, arrow wounds, 54
1868 — Jesse Chisholm, cholera, 62
1884 — John S. Chisum, neck tumor, 60
1885 — Richard King, stomach cancer, 60
1895 — Pete Kitchen, 76
1896 — Marquis de Mores, killed by Touareg natives, 38
1900 — Shanghai Pierce, 66
1904 — Dan Waggoner, 76
1906 — Walter Vail, streetcar accident, 54
1907 — Henry Hooker, 79
1911 — W.C. Greene, runaway team, 57
        Clem Rogers, 72
1913 — Pierre Wibeaux, liver cancer, 54
1914 — Pink Higgins, heart attack, 62
1915 — Joseph G. McCoy, 77
1916 — Henry Miller, 89
1917 — William F. Cody, 70
1918 — Granville Stuart, 84
1919 — Theodore Roosevelt, 61
        Ora Haley, 74
        C.C. Slaughter, 81
1920 — George W. Littlefield, 78
        Conrad Kohrs, 84
1922 — John Slaughter, 80
        Burke Burnett, 73
1925 — Henrietta King, 92
1926 — Charles Russell, heart attack, 62
1928 — Charles Siringo, 73
1929 — Charles Goodnight, 93
1932 — Bill Pickett, kicked by a horse, 62
        Robert Kleberg, 78
1933 — John B. Kendrick, 86
1934 — W.T. Waggoner, 82
1935 — Andy Adams, 76
1939 — Murdo Mackenzie, 89
1956 — Burt Mossman, 89

BIBLIOGRAPHY

Adams, Andy. *The Log of a Cowboy: A Narrative of the Old Trail Days.* Garden City, New York: Doubleday & Company, Inc., 1964.

Athearn, Robert G. *The Coloradans.* Albuquerque: University of New Mexico Press, 1976.

Atherton, Lewis. *The Cattle Kings.* Bloomington: Indiana University Press, 1961.

Bailey, Lynn R. *"We'll All Wear Silk Hats."* Tucson, Arizona: Westernlore Press, 1994.

Bancroft, H. H. *History of Texas and the North Mexican States.* San Francisco: History Company, 1890.

Barber, Floyd R., and Dan W. Martin. *Idaho in the Pacific Northwest.* Caldwell, Idaho: The Caxton Printers, Ltd., 1956.

Bard, Floyd C., told to Agnes Wright Spring. *Horse Wrangler: Sixty Years in the Saddle in Wyoming and Montana.* Norman: University of Oklahoma Press, 1960.

Barrows, John R. *U-bet: A Greenhorn in Old Montana.* Lincoln: University of Nebraska Press, 1990.

Bassford, Forrest. *Wyoming Hereford Ranch, 1883-1983: A Century of Endurance.* N. p., 1983.

Beattie, George Williana, and Helen Pruitt. *Heritage of the Valley:San Bernardino's First Century.* Oakland, California: Biobooks, 1951.

Bechdolt, Frederick R. *Riders of the San Pedro.* New York: A.L. Burt Company, Publishers, 1931.

Bentley, Henry Lewis. *Cattle Ranges of the Southwest: A History of Exhaustion of Pasturage and Suggestions for Its Restoration.* United States Department of Agriculture, Farmers Bulletin No. 72. Washington, D.C.: 1898.

Benton, Minnie King. *Boomtown: A Portrait of Burkburnett.* Quanah and Wichita Falls, Texas: Nortex Offset Publications, Inc., 1972.

Black, Roe R. *The Horseshoe-Bar Ranch: Remembering a Prairie Childhood.* Lincoln, Nebraska: Media Publishing, 1985.

Blasingame, Ike. *Dakota Cowboy: My Life in the Old Days.* Lincoln: University of Nebraska Press, 1964.

Borne, Lawrence R. *Dude Ranching: A Complete History.* Albuquerque: University of New Mexico Press, 1983.

Brackett, Robert W. *A History of the Ranches of San Diego County, California.* San Diego: Union Title Insurance and Trust Company, 1939.

Brisbin, Gen. James S. *The Beef Bonanza; or, How to Get Rich on the Plains.* Norman: University of Oklahoma Press, 1959 [1881].

Brooks, Chester L., and Ray H. Mattison. *Theodore Roosevelt and the Dakota Badlands.* Washington, D.C.: National Park Service, 1958.

Brooks, Connie, *The Last Cowboys: Closing the Open Range in Southeastern New Mexico, 1890s-1920s.* Albuquerque: University of New Mexico Press, 1993.

Bryant, Tom, and Joel Bernstein. *A Taste of Ranching: Cooks & Cowboys.* Albuquerque, New Mexico: Border Books, 1993.

Burdick, Usher Lloyd. *Marquis de Mores at War in the Badlands.* Fargo, North Dakota: N. p., 1930.

Burns, Robert Homer, Andrew Springs Gillespie, and Willing Gay Richardson. *Wyoming's Pioneer Ranches.* Laramie: Top-of-the-World Press, 1955.

Burroughs, John Rolfe. *Where the Old West Stayed Young.* New York: Bonanza Books, 1962.

Carlock, Robert H. *The Hashknife: The Early Days of the Aztec Land and Cattle Company, Limited.* Tucson: Westernlore Press, 1994.

Clark, Mary Whatley. *John Chisum, Jinglebob King of the Pecos.* Austin: Eakin Press.

Clarke, Norm. *Tracing Terry Trails: A Chronological History.* N. p., 1982.

Clayton, Lawrence. *Historic Ranches of Texas.* Austin: University of Texas Press, 1993.

Cleland, Robert Glass. *The Cattle on a Thousand Hills: Southern California, 1850-1880.* San Marino, California: The Huntington Library, 1951.

———. *The Irvine Ranch of Orange County, 1810-1950.* San Marino, California: The Huntington Library, 1952.

Collings, Ellsworth, and Alma Miller England. *The 101 Ranch.* Norman: University of Oklahoma Press, 1938.

Conner, Palmer. *The Romance of the Ranchos.* Los Angeles: Title Insurance and Trust Company, 1939.

Coolidge, Dane. *Arizona Cowboys.* Tucson, Arizona: The University of Arizona Press, 1938.

Cowan, Robert Elsworth (Bud). *Range Rider.* Garden City, N.Y.: Doubleday, Doran & Co., Inc., 1930.

Cowan, Robert G. *Ranchos of California.* Fresno, California: Academy Library Guild, 1956.

Cox, James. *Historical and Biographical Record of the Cattle Industry and the Cattlemen: Texas and Adjacent Territory.* St. Louis: Woodward & Tiemam, 1895.

Cypher, John. *Bob Kleberg and the King Ranch: A Worldwide Sea of Grass.* Austin: University of Texas Press, 1995.

Dale, Edward Everett. *The Range Cattle Industry: Ranching on the Great Plains from 1865 to 1925.* Norman: University of Oklahoma Press, 1930.

David, Robert B. *Malcolm Campbell, Sheriff.* Casper, Wyoming: Wyomingana, Inc., 1932.

Dobie, J. Frank. *Cow People.* Boston: Little, Brown and Company, 1964.

Douglas, C.L. *Cattle Kings of Texas.* Austin: State House Press, 1989.

Dresden, Donald. *The Marquis de Mores, Emperor of the Bad Lands.* Norman: University of Oklahoma Press, 1970.

Duke, Cordia Sloan, and Joe B. Frantz. *6,000 Miles of Fence.* Austin: University of Texas Press, 1961.

Dykstra, Robert R. *The Cattle Towns.* New York: Atheneum, 1970.

Edgar, Bob, and Jack Turnell. *Brand of a Legend.* N.p., 1978.

Eggen, John E. *Cowboys.* West Chester, Pennsylvania: Schiffer Publishing, Ltd., 1992.

————. *"The West That Was."* West Chester, Pennsylvania: Schiffer Publishing, Ltd., 1991.

Ellis, George F. *Bell Ranch As I Knew It.* Kansas City, Missouri: The Lowell Press, 1973.

Ellis, Martha Downer. *Bell Ranch, Places and People.* Clarendon, Texas: Clarendon Press, 1963.

————. *Bell Ranch Recollection and Memories.* Conchas Dam, New Mexico: Ellis Book Company, 1985.

Ellis, Mattie, and Mark Wood. *Bell Ranch Wagon Work.* Conchas Dam, New Mexico: Ellis Book Compnay, 1984.

Emmett, Chris. *Shanghai Pierce: A Fair Likeness.* Norman: University of Oklahoma Press, 1953.

Erwin, Allen A. *The Southwest of John H. Slaughter, 1841-1922.* Glendale, California: Arthur H. Clark Company, 1965.

Faulkner, Virginia, ed. *Roundup: A Nebraska Reader.* Lincoln: University of Nebraska Press, 1957.

Fleming, Elvis E., and Ernestine Chesser Williams. *Treasures of History II, Chaves County Vignettes.* Roswell, New Mexico: Chaves County Historical Society, 1991.

Fontana, Bernard L., intro. *An Englishman's Arizona, The Ranching Letters of Herbert R. Hislop, 1876-1878.* Tucson: The Overland Press, 1965.

French, Giles. *Cattle Country of Peter French.* Portland, Oregon: Binfords & Mort, Publishers, 1964.

Frink, Maurice. *The Boulder Story.* Boulder, Colorado: Pruett Press, Inc., 1965.

Frissell, Toni, and Holland McCombs. *The King Ranch, 1939-1944.* Dobbs Ferry, New York: Morgan & Morgan, 1975.

Goodwyn, Frank. *Life on the King Ranch.* New York: Thomas Y. Crowell Company, 1951.

Goplen, Arnold O. *The Career of the Marquis de Mores in the Badlands of North Dakota.* Bismarck, N.D.: State Historical Society of North Dakota, 1979.

Graham, Joe S. *El Rancho in South Texas: Continuity and Change From 1750.* Denton, Texas: University of North Texas Press, 1994.

Gregg, Andrew K. *New Mexico in the Nineteenth Century: A Pictorial History.* Albuquerque: University of New Mexico Press, 1968.

Grosskopf, Linda, with Rick Newby. *On Flatwillow Creek: The Story of Montana's N Bar Ranch.* Los Alamos, New Mexico: Exceptional Books, Ltd., 1991.

Haley, J. Evetts. *Charles Goodnight, Cowman and Plainsman.* Norman: University of Oklahoma Press, 1949.

————. *George W. Littlefield, Texan.* Norman: University of Oklahoma Press, 1943.

————. *The XIT Ranch of Texas.* Norman: University of Oklahoma Press, 1953.

Hamlin, James D., with J. Evetts Haley and William Curry Holden. *The Flamboyant Judge.* Canyon, Texas: Palo Duro Press, 1972.

Hamner, Laura V. *Short Grass and Longhorns.* Norman: University of Oklahoma Press, 1943.

Hanes, Colonel Bailey C. *Bill Pickett, Bulldogger: The Biography of a Black Cowboy.* Norman: University of Oklahoma Press, 1977.

Hastings, Frank S. *A Ranchman's Recollections.* Austin: The Texas State Historical Association, 1985.

Hill, Joseph J. *The History of Warner's Ranch and Its Environs.* Los Angeles: Privately printed, 1927.

Hinshaw, Gil. *Lea, New Mexico's Last Frontier.* Hobbs, New Mexico: The Hobbs *Daily News-Sun,* 1976.

Hinton, Harwood P., intro. *History of the Cattlemen of Texas.* Austin: Texas State Historical Association, 1991.

Holden, W. C. *Rollie Burns, or An Account of the Ranching Industry on the South Plains.* College Station: Texas A&M University Press, 1982.

———. *The Espuela Land and Cattle Company.* N.p., 1970.

Hughes, Stella. *Hashknife Cowboy: Recollections of Mack Hughes.* Tucson: The University of Arizona Press, 1984.

Hunt, Frazier. *Cap Mossman: Last of the Great Cowmen.* New York: Hastings House, 1951.

Hunt, Rockwell D. *John Bidwell, Prince of California Pioneers.* Caldwell, Idaho: The Caxton Printers, Ltd., 1942.

Hunter, J. Marvin, ed. *The Trail Drivers of Texas.* Austin: University of Texas Press, 1984.

Hutchinson, C. Alan. *Frontier Settlement in Mexican California.* New Haven and London: Yale University Press, 1969.

Johnson, Cecil. *Guts, Legendary Black Rodeo Cowboy Bill Pickett.* Fort Worth, Texas: The Summit Group, 1994.

Johnson, Emily. *The White House Ranch.* N.p.: O'Brien Printing and Lithographic Press, 1972.

Johnson, Frank W., with Eugene C. Barker and Ernest William Winkler. *A History of Texas and Texans.* Vol. I. Chicago and New York: The American Historical Society, 1914.

Jones, Helen Carey, compiler. *Custer County Area History: As We Recall, A Centennial History of Custer County, Montana.* Dallas, Texas: Curtis Media Corporation, 1990.

Jordan, Terry G. *North American Cattle-Ranching Frontiers: Origins, Diffusions, and Differentiation.* Albuquerque: University of New Mexico Press, 1993.

Kelton, Steve. *Renderbrook: A Century Under the Spade Brand.* Fort Worth: Texas Christian University Press, 1989.

Kennedy, Michael S. *Cowboys and Cattlemen.* New York: Hastings House, Publishers, 1964.

Kerr, W.G. *Scottish Capital on the American Credit Frontier.* Austin: Texas State Historical Association, 1976.

Ketchum, Richard M. *Will Rogers, His Life and Times.* New York: American Heritage Publishing Company, Inc., 1973.

King, Bucky. *The Empire Builders: The Development of Kendrick Cattle Company.* N.p., 1992.

Kohrs, Conrad. *Conrad Kohrs: An Autobiography.* Deer Lodge, Montana: Platen Press, 1977.

Lamar, Howard R., ed. *The Reader's Encyclopedia of the American West.* New York: Thomas Y. Crowell Company, 1977.

Lanning, Jim and Judy, eds. *Texas Cowboys, Memories of the Early Days.* College Station: Texas A&M University Press, 1984.

Lea, Tom. *The King Ranch.* 2 vols. Boston: Little, Brown and Company, 1957.

Lincoln, John. *Rich Grass and Sweet Water: Ranch Life with the Koch Matador Cattle Company.* College Station: Texas A&M University Press, 1989.

Lockwood, Frank C. *Arizona Characters.* Los Angeles: The Times-Mirror Press, 1928.

McAfee, W.R. *The Cattlemen.* Fort Davis: Davis Mountain Press, 1989.

McCarty, John L. *Maverick Town: The Story of Old Tascosa.* Norman: University of Oklahoma Press, 1946.

McCoy, Jeseph G. *Historic Sketches of the Cattle Trade of the West and Southwest.* Lincoln: University of Nebraska Press, 1985.

McCullough, David. *Mornings on Horseback.* New York: Simon and Schuster, 1981.

McNeill, J.C. ("Cap"), III. *The McNeills' SR Ranch, 100 Years in Blanco Canyon.* College Station: Texas A&M University Press, 1988.

Mercer, A.S. *The Banditti of the Plains.* Norman: University of Oklahoma Press, 1954.

Metz, Leon C. *Pat Garrett.* Norman: University of Oklahoma Press, 1974.

Miller, Nyle H., Edgar Longsdorf, and Robert W. Richmond. *Kansas in Newspapers.* Topeka: Kansas State Historical Society, 1963.

Miner, Craig. *West of Wichita: Settling the High Plains of Kansas, 1865-1890.* Topeka: University Press of Kansas, 1986.

Mokler, A.J. *History of Natrona County.* Chicago: Lakeside Press, 1923.

Morrison, Lorin L. *Warner, the Man and the Ranch.* Los Angeles: Lorin L. Morrison, 1962.

Murrah, David J. *C. C. Slaughter, Rancher, Banker, Baptist.* Austin: University of Texas Press, 1981.

———. *The Pitchfork Land and Cattle Company: The First Century.* Lubbock, Texas: PrinTech, Texas Tech University, 1983.

Nelson, Edna Deu Pree. *The California Dons.* New York: Appleton-Century-Crofts, Inc., 1962.

Oliver, Herman. *Gold and Cattle Country.* Portland, Oregon: Binfords & Mort Publishers, 1962.

Olson, James C. *History of Nebraska.* Lincoln: University of Nebraska Press, 1955.

Paddock, Capt. B.B. *History and Biographical Record of North and West Texas.* 2 vols. Chicago and New York: The Lewis Publishing Co., 1906.

Parsons, Jack, and Michael Earney. *Land and Cattle, Conversations with Joe Pankey, a New Mexico Rancher.* Albuquerque: University of New Mexico Press, 1978.

Paul, Virginia. *This was Cattle Ranching, Yesterday and Today.* New York: Bonanza Books, 1973.

Peake, Ora Brooks. *The Colorado Range Cattle Industry.* Glendale, California: The Arthur H. Clark Company, 1937.

Pearce, W.M. *The Matador Land and Cattle Company.* Norman: University of Oklahoma Press, 1964.

Pitt, Leonard. *The Decline of the Californios.* Berkeley and Los Angeles: University of California Press, 1966.

Pool, William C. *A Historical Atlas of Texas.* Austin: The Encino Press, 1975.

Porter, Willard H. *Who's Who in Rodeo*. Oklahoma City: Powder River Book Company, 1982.

Pringle, Henry F. *Theodore Roosevelt: A Biography*. New York: Harcourt, Brace and Company, 1931.

Procter, Gil. *The Trails of Pete Kitchen*. Tucson: Dale Stuart King, Publisher, 1964.

Propst, Nell Brown. *The South Platte Trail, Story of Colorado's Forgotten People*. Boulder, Colorado: Pruett Publishing Co., 1979.

Rathjen, Frederick W. *The Texas Panhandle Frontier*. Austin: University of Texas Press, 1973.

Remley, David. *Bell Ranch: Cattle Ranching in the Southwest, 1824-1947*. Albuquerque: University of New Mexico Press, 1993.

Robertson, Pauline Durrett, and R. L. Robertson. *Panhandle Pilgrimage*. Canyon, Texas: Staked Plains Press, Inc., 1976.

Robinson, W.W. *Ranches Become Cities*. Pasadena, California: San Pasqual Press, 1939.

Rollinson, John K. *Wyoming Cattle Trails*. Caldwell, Idaho: The Caxton Printers, Ltd., 1948.

Roosevelt, Theodore. *The Autobiography of Theodore Roosevelt*. New York: Charles Scribner's Sons, 1958.

―――. *Ranch Life in the Far West*. Flagstaff, Arizona: Northland Press, 1985 [1888].

Russell, Don. *The Lives and Legends of Buffalo Bill*. Norman: University of Oklahoma Press, 1960.

Russell, Jim. *Bob Fudge, Texas Trail Driver, Montana-Wyoming Cowboy, 1862-1933*. Aberdeen, South Dakota: North Plains Press, 1981.

Sandoval, Judith Hancock, with T. A. Larson and Robert Roripaugh. *Historic Ranches of Wyoming*. Casper, Wyoming: Nicolaysen Art Museum and Mountain States Lithography Company, 1986.

Sandoz, Mari. *The Cattlemen, From the Rio Grande Across the Far Marias*. New York: Hastings House, Publishers, 1958.

Savage, William W., Jr. *The Cherokee Strip Live Stock Association*. Norman: University of Oklahoma Press, 1973.

Schell, Herbert S. *History of South Dakota*. Lincoln: University of Nebraska Press, 1961.

Schlebecker, John T. *Cattle Raising on the Plains, 1900-1961*. Lincoln: University of Nebraska Press, 1963.

Sheffly, Lester Fields. *The Francklyn Land and Cattle Company: A Panhandle Enterprise, 1882-1957*. Austin: The University of Texas, 1963.

Shirley, Glenn. *Pawnee Bill: A Biography of Major Gordon W. Lillie*. Stillwater, Oklahoma: Western Publications, 1993.

Siberts, Bruce. *Nothing but Prairie and Sky: Life on the Dakota Range in the Early Days*. Norman: University of Oklahoma Press, 1954.

Simmons, Marc. *Ranchers, Ramblers and Renegades: True Tales of Territorial New Mexico*. Santa Fe: Ancient City Press, 1984.

Sinclair, John L. *Cowboy Riding Country*. Albuquerque: University of New Mexico Press, 1982.

Skaggs, Jimmy M. *The Cattle-Trailing Industry, Between Supply and Demand, 1866-1890.* Norman: University of Oklahoma Press, 1991.

Smith, Earl R. *The Westfall Country.* New York: Exposition Press, 1963.

Smith, Erwin E., and J. Evetts Haley. *Life on the Texas Range.* Austin: University of Texas Press, 1952.

Sonnichsen, C.L. *Tularosa, Last of the Frontier West.* Albuquerque: University of New Mexico Press, 1980.

Steber, Rick. *Cowboys.* Prineville, Oregon: Bonanza Publishing, 1988.

Steiner, Stan. *The Ranchers: A Book of Generations.* New York: Alfred A. Knopf, 1980.

Stephenson, Terry E. *Don Bernardo Yorba.* Los Angeles: Glen Dawson, 1941.

Stewart, Janet Ann. *Arizona Ranch Houses, Southern Territorial Styles, 1867-1900.* Tucson: University of Arizona Press and Arizona Historical Society, 1974.

Sullivan, Dulcie. *The LS Brand.* Austin: The University of Texas Press, 1968.

Tanner, Ogden. *The Ranchers.* Alexandria, Virginia: Time-Life Books, 1977.

Thrapp, Dan L. *Encyclopedia of Frontier Biography.* 3 vols. Lincoln: University of Nebraska Press, 1988.

Tinsley, Jim Bob. *The Hash Knife Brand.* Gainesville: University Press of Florida, 1993.

Towne, Charles Wayland, & Edward Norris Wentworth. *Cattle & Men.* Norman: University of Oklahoma Press, 1955.

Traywick, Ben T. *The Law and John Slaughter.* Tombstone: Red Marie's, 1983.

Treadwell, Edward F. *The Cattle King.* New York: The Macmillan Company, 1931.

Trenholm, Virginia Cole. *Footprints on the Frontier.* Douglas, Wyoming: Douglas Enterprise Co., 1945.

Trimble, Marshall. *Arizona: A Panoramic History of a Frontier State.* Garden City, N.Y.: Doubleday & Company, Inc., 1977.

———. *Arizona Adventure!* Phoenix: Golden West Publishers, 1982.

Utley, Robert M. *High Moon in Lincoln: Violence on the American Frontier.* Albuquerque: University of New Mexico Press, 1987.

Vanderbilt, Cornelius, Jr. *Ranches and Ranch Life in America.* New York: Crown Publishers, Inc., 1968.

Viola, Herman J., and Sarah Loomis Wilson, eds. *Texas Ranchman: The Memoirs of John A. Loomis.* Chadron, Nebraska: The Fur Press, 1982.

Walker, Peggy. *George Humphreys, Cowboy and Lawman.* Burnet, Texas: Eakin Publications, 1978.

Ward, Delbert R. *Great Ranches of the United States.* San Antonio: Ganado Press, 1993.

Welsh, Donald H. *Pierre Wibeaux, Cattle King.* Bismarck, N.D.: The State Historical Society of North Dakota, 1953.

Wheels Across Montana's Prairie. Terry, Montana: Prairie County Historical Socie ty, 1974.

White, Benton R. *The Forgotten Cattle King.* College Station: Texas A&M Unive· sity Press, 1986.

Wibeaux County Diamond Jubilee and Montana Centennial, 1989. N.p., 1989.

Williams, J.W. *The Big Ranch Country.* Burnet, Texas: Eakin Publications, 1984

Willson, Roscoe G. *Pioneer Cattlemen of Arizona.* Vol. I. Phoenix: Valley National Bank, 1951.

Wilson, Iris Highbie. *Willliam Wolfskill, 1798-1866: Frontier Trapper to California Ranchero.* Glendale, California: The Arthur H. Clark Company, 1965.

Worcester, Don. *The Chisholm Trail, High Road of the Cattle Kingdom.* Lincoln: University of Nebraska Press, 1980.

———. *The Texas Longhorn, Relic of the Past, Asset for the Future.* College Station: Texas A&M University Press, 1987.

Wyoming: A Guide to Historic Sites. Basin, Wyoming: Big Horn Publishers, 1988.

Yagoda, Ben. *Will Rogers: A Biography.* New York: Alfred A. Knopf, 1994.

Yeats, E.L., and Hooper Shelton. *The 18 Ranch: Colorado Cattle Co., 1881-1973.* N.p.: Feather Press, n.d.

Yost, Nellie Snyder. *Buffalo Bill: His Family, Friends, Fame, Failures, and Fortunes.* Chicago: Sage Books, 1979.

———, ed. *The Recollections of Ed Lemmon, 1857-1946.* Lincoln: University of Nebraska Press, 1969.

Young, James A., and B. Abbott Sparks. *Cattle in the Cold Desert.* Logan: Utah University Press, 1985.

Articles

Borne, Lawrence R. "Dude Ranching in the Rockies." Montana, *The Magazine of Western History,* vol. XXXVIII, no. 3, (Summer 1988.)

Box, Thadis W. "Range Deterioration in West Texas," *Southwestern Historical Quarterly,* vol. LXXI, no. 1 (July 1967.)

Briggs, Harold E. "Ranching and Stock Raising in the Territory of Dakota." *South Dakota Historical Collections,* vol. XIV (1928).

Broyles, William, Jr. "The Last Empire." *Texas Monthly,* October 1980.

Burns, Robert H. "The Newman Brothers, Forgotten Cattle Kings of the Northern Plains." *Montana, The Magazine of Western History,* vol. XI, no. 4 (Autumn 1961).

———. "The Newman Ranches: Pioneer Cattle Ranches of the West." *Nebraska History,* vol. XXXIV (March 1953).

Burton, Harley T. "History of the J.A. Ranch." *Southwestern Historical Quarterly,* Vol. XXXI (October 1927).

Cunningham, Robert. "Pennsylvania's Camerons in Bloodshed on the Border." *Journal of Arizona History,* vol. 89, no. 3 (1985).

Davis, Elmer W. "Where the Ancient and Modern Meet — Tanque Verde Ranch." *Progressive Arizona and the Great Southwest,* vol. VII (September 1928).

Fleming, Elvis. "Hagerman Mansion Remains a County Landmark." *Roswell Daily Record,* September 12, 1988.

Ford, Lee M. "Bob Ford, Sun River Cowman." *Montana, The Magazine of Western History,* vol. IX, no. 1 (Winter 1959).

Forrest, Earle R. "The Fabulous Sierra Bonita." *Journal of Arizona History,* vol. VI, no. 6 (Autumn 1965).

Frantz, Joe B. "Texas' Largest Ranch — In Montana." *Montana, The Magazine of Western History,* vol. XI, no. 4, (Autumn 1961).

Fudge, Bob. "Long Trail From Texas." *Montana, The Magazine of Western History*, vol. XII, no. 3 (Summer 1962).

Gard, Wayne. "The Impact of the Cattle Trails." *Southwestern Historical Quarterly*, vol. LXXI, no. 1 (July 1967).

Gill, Larry. "From Butcher to Beef King." *Montana, The Magazine of Western History*, vol. VIII, no. 2 (Spring 1958).

Guthrie, W.E. "The Open Range Cattle Business in Wyoming." *Annals of Wyoming*, vol. V (July 1927).

Haskett, Bert. "Early History of the Cattle Industry in Arizona." *Arizona Historical Review*, vol. VI (October 1935).

Hill, Gertrude. "Henry Clay Hooker: King of the Sierra Bonita." *Arizoniana*, vol. II, no. 4 (Winter 1961).

Hinton, Harwood P., Jr. "John Simpson Chisum, 1877-84." *New Mexico Historical Review*, vol. XXXI, no. 3 (July 1956) and no. 4 (October 1956) and vol. XXXII, no. 1 (January 1957).

Holden, W. C. "The Problem of Stealing on the Spur Ranch." *West Texas Historical Association Year Book*, vol. VIII (June 1932).

———, ed. "A Spur Ranch Diary, 1887." *West Texas Historical Association Year Book*, vol. VII (June 1931).

Kelton, Elmer. "Rancher Use of Range Resource." *The Roundup Quarterly*, vol. V, no. 3 (Spring 1993).

Kittredge, William, and Steven M. Krauzer. "`Mr. Montana' Revised." *Montana, The Magazine of Western History*, vol. XXVI, no. 4 (Autumn 1986).

Lockwood, Frank C. "Pete Kitchen: Arizona Pioneer Rifleman and Ranchman." *Arizona Historical Review*, vol. I (April 1928).

Love, Clara M. "History of the Cattle Industry in the Southwest." *Southwestern Historical Quarterly*, vol. XIX (April 1916), and vol. XX (July 1916).

MacMillan, D. "The Gilded Age and Montana's DHS Ranch." *Montana, The Magazine of Western History*, vol. XX, no. 2 (Spring 1970).

Oliphant, J. Orin. "The Cattle Trade from the Far Northwest to Montana." *Agricultural History*, vol. XI (April 1932).

Perkins, Doug. "In Search of the Origins of the Texas Cattle Industry." *The Cattleman*, March 1977.

Randall, L.W. (Gay). "The Man Who Put the Dude in Dude Ranching." *Montana, The Magazine of Western History*, vol. X, no. 3 (Summer 1960).

Reese, William S. "Granville Stuart of the DHS Ranch, 1879-1887." *Montana, The Magazine of Western History*, vol. XXI, no. 3 (Summer 1981).

Remley, David. "`To Struggle Against an Adverse Fate': Granville Stuart, Cowman." *Montana*, vol. XXI, no. 3 (Summer 1981).

Richardson, Ernest M. "Moreton Frewen: Cattle King With Monocle." *Montana, The Magazine of Western History*, vol. XI, no. 4 (Autumn 1961).

Rylander, Dorothy. "The Economic Phase of the Ranching Industry on the Spur Ranch." *West Texas Historical Association Year Book*, vol. VII (June 1931).

Skaggs, Jimmy M. "John Thomas Lytle: Cattle Baron." *Southwestern Historical Quarterly*, vol. LXXI, no. 1 (July 1967).

Smith, Dean. "The Babbitts, Arizona Cattlemen for More Than 75 Years." *Arizona Cattlelog*, November 1964.

Snoke, Elizabeth R. "Pete Kitchen, Arizona Pioneer." *Arizona and the West,* vol. XXI, no. 3 (July 1963).

"The Spade and Renderbrook Ranch." *The Great Southwest,* vol. I, no. 2 (January 25, 1940).

Traywick, Ben. "Famous Lawman's Ranch on Historic Land." *The Tombstone Epitaph,* March 1993.

Warren, Particia Neal. "Saga of an American Ranch." Reader's Digest, January 1982.

Wheeler, David L. "The Blizzard of 1882 and Its Effect on the Range Cattle Industry in the Southern Plains." *Southwestern Historical Quarterly,* vol. XCIV, no. 3 (January 1991).

——— "Winter on the Cattle Range, Western Kansas, 1884-1886." *Kansas History,* vol. XV, no. 1 (Spring 1992).

Wilson, James A. "West Texas Influence on the Early Cattle Industry of Arizona." *Southwestern Historical Quarterly,* vol. LXXI, no. 1 (July 1967).

Wood, A.B. "The Coad Brothers; Panhandle Cattle Kings." *Nebraska History,* vol. XIX (January 1938).

Wood, Charles L. "Cattlemen, Railroads, and the Origin of the Kansas Livestock Association." *The Kansas Historical Quarterly,* vol. XLIII, no. 2 (Summer 1977).

Young, James A. "The Dreaded Winter of '89." *Nevada,* Jan./Feb. 1984.

Newspapers

Arizona Daily Star   Holbrook *Tribune-News*
Arizona Republic   Lampasas *Daily Leader*
Carlsbad *Current*   Tombstone *Epitaph*
Dallas *Morning News*   Tucson *Daily Citizen*

Miscellaneous

Arizona Historical Society Files:
   Colin Cameron   William C. Greene
   Burt Mossman   Henry C. Hooker
   Sierra Bonita Ranch   Empire Ranch
   Tanque Verde Ranch   Emilio Carrillo
   Aztec Land & Cattle Company   San Bernardino Ranch
                                  Babbitt Brothers

Arthur Bowring Sandhills Ranch. Pamphlet prepared by the Nebraska Game and Parks Commission.

Bayer, Vicky. "History of the Z Bar Ranch." Made available by the Chase County Historical Society, Cottonwood Falls, Kansas.

The Bradford Brinton Memorial. Pamphlet prepared by the Bradford Brinton Memorial, Big Horn, Wyoming.

"A Brief History of the Cattle Industry in Texas." Prepared by Texas and Southwestern Cattle Raisers Association, *The Cattleman Magazine.*

A Brief History of the Spade Ranches, Established in 1889. Promotional literature.

Dowell, Gregory Paul. "History of the Empire Ranch." Unpublished master's thesis at the University of Arizona, Tucson.

Empire Ranch Files. Bureau of Land Management Office. Tucson, Arizona.

Empire-Cienega Resource Conservation Area. Pamphlet prepared by Tucson Resource Area.

"The Goose Egg Ranch House." Maunscript prepared and printed by Wyoming Field Science Foundation, Casper.

McClure, Charles Boone. "A History of Randall County and the T Anchor Ranch." Master's thesis in the Southwest Collection at Texas Tech University, Lubbock. 1930.

McKenny, Kregg. "A History of the Elwood Ranches." Research paper in the Southwest Collection at Texas Tech University, Lubbock. 1977.

Oliver Lee Memorial State Park. Pamphlet prepared by New Mexico State Parks.

Pierre Wibeaux House Museum. Pamphlet prepared by Wibeaux County Museum.

"Ranches and Ranching." File, Historical Research Collection, Wyoming State Museum, Cheyenne.

Rowlinson, Don D. "An English Settlement in Sheridan County, Kansas." 1988 paper developed by Don D. Rowlinson, curator of Cottonwood Ranch.

"Self Guided Tour." Typescript prepared by John Slaughter Ranch, San Bernardino National Historic Landmark.

The Spade Ranch. Promotional literature.

Sun Ranch Since 1970. Typescript prepared and printed by Sun Ranch, Alcova, Wyoming.

Trail End State Historic Site. Pamphlet prepared by Wyoming State Parks & Historic Sites, Wyoming Department of Commerce, Cheyenne.

Warnke, Tom. "Profile of John B. Kendrick and the OW Ranch." Typescript provided by Trail End State Historic Site, Sheridan, Wyoming.

White House Ranch Historic Site. Pamphlet prepared by Colorado Springs Park and Recreation Department.

INDEX

CK Ranch, 216-219
Clanton, Ike, 251
Clark, William, 202
Clay, John, 187, 309
Clay Springs Cattle Company, 170
Clement, Charlie, 229
Cleveland, Grover, 123
CO Bar, 274
Cochise, 247
Cody, Buffalo Bill, xii, 120, 128, 141-
  150, 205
  Lulu, 141, 148, 149-150
  William F., 334
Cody, Wyoming, 148
Coffeldt, Matt, 101
Coffeyville, Kansas, 116
Collins, Herb, 37
Collinson, Frank, 46
Colorado City, Colorado, 170
Colorado City, Texas, 84
Comanche Cattle Pool, 135
Comes a Horseman, 333
Comstock, Mrs., 154
  William G., 154
Comstock Lode, 247
Continental Cattle Company, 268
Continental Divide, 213
Continental Land and Cattle
    Company, 268
Converse, James, 245
Converse Cattle Company, 176
Cooper, Gary, 333
  Mary Ellen, 231
Cooweescoowee, Oklahoma, 113, 115
Cornell University, 97
Cottonwood Ranch, 136-139
Couts, J. R., 267, 268
Couts and Simpson, 267-268
Cowboy, 333
Cowboys, The, 333
Cowboy Strike of 1883, 79
Cow Puncher, 63
Crazy Woman Creek, 200
Crook, George, 248, 286
Crow Creek, 188
Crowell, Henry Parsons, 190
Crow Reservation, 178
Culpepper Cattle Company, 332

Cummings, Daniel, 313
Curtis Act, 118
Custer Trail Cattle Company, 292
D
Dakota Division, 43
Dalton brothers, 117
Danks, Clayton, 183
  Jimmie, 182
Dart, Isom, 169
Davis, Andrew J., 221, 222
  "Diamondfield Jack," 313-314
  Edwin, 221, 222, 225
  George, 134, 136
  Harry, 180
Davis, Hauser and Company, 221
Dawson, John, 38
de Baca, Luis Maria Cabeza, 105
Decatur, Texas, 64, 65
Deer Creek Ranch, 133
Deer Lodge, Montana, 213, 215
Deer Lodge Valley, 221
Dempsey, Jack, 120
Denby, Al, 225, 227, 228
Denver, Colorado, 160
Desert Land Act, 206
Devil's Gate, 203, 205, 206
DHS, 221-225
Diamond A Cattle Company, 97
Dillon, John L., 256
Dixon Creek, 69, 70
Dog Canyon, 99-104
Dog Canyon Ranch, 99-104
Dog Iron Ranch, 113, 115-118, 326
Doheny, Ed, 24
Doolin, Bill, 102
Double H Ranch, 321-323
Double O Ranch, 149
Doughty, James, 7
Douglas, Kirk, 333
Douglas, Arizona, 286
Dow, Will, 297, 298, 301
Downing, Bill, 307, 308
Downs, Billy, 223
Drought, Harry P., 44
Dudley, Nathan, 94
Duel in the Sun, 332
Dunnie, 35
Dusky Demon (*see* Pickett, Bill)
Dyer, Leigh, 79, 163